Lecture Notes in Mathematics

Edited by A. Dold, B. Eckmann and F. Takens

Subseries: Instituto de Matemática Pura e Aplicada, Rio de Janeiro
Adviser: C. Camacho

1430

W. Bruns   A. Simis   (Eds.)

Commutative Algebra

Proceedings of a Workshop held in
Salvador, Brazil, Aug. 8–17, 1988

Springer-Verlag

Berlin Heidelberg New York London
Paris Tokyo Hong Kong Barcelona

Editors

Winfried Bruns
Universität Osnabrück—Standort Vechta, Fachbereich Mathematik
Driverstraße 22, 2848 Vechta, FRG

Aron Simis
Universidade Federal da Bahia, Departamento de Matemática
Av. Ademar de Barros, Campus de Ondina
40210 Salvador, Bahia, Brazil

Mathematics Subject Classification (1980): 13-04, 13-06, 13C05, 13E13,
13C15, 13D25, 13E15, 13F15,
13H10, 13H15

ISBN 3-540-52745-1 Springer-Verlag Berlin Heidelberg New York
ISBN 0-387-52745-1 Springer-Verlag New York Berlin Heidelberg

Printing and binding: Druckhaus Beltz, Hemsbach/Bergstr.
2146/3140-543210 – Printed on acid-free paper

Introduction

This volume contains the proceedings of the Workshop in Commutative Algebra, held at Salvador (Brazil) on August 8–17, 1988. A few invited papers were included which were not presented in the Workshop. They are, nevertheless, very much in the spirit of the discussions held in the meeting.

The topics in the Workshop ranged from special algebras (Rees, symmetric, symbolic, Hodge) through linkage and residual intersections to free resolutions and Gröbner bases. Beside these, topics from other subjects were presented such as the theory of maximal Cohen-Macaulay modules, the Tate-Shafarevich group of elliptic curves, the number of rational points on an elliptic curve, the modular representations of the Galois group and the Hilbert scheme of elliptic quartics. Their contents were not included in the present volume because the respective speakers felt that the subject had been or was to be published somewhere else. Other, more informal, lectures were presented at the meeting that are not included here either for similar reasons.

The meeting took place at the Federal University of Bahia, under partial support of CNPq and FINEP to whom we express our gratitude. For the practical success of the Workshop we owe an immense debt to Aron Simis' wife, Lu Miranda, whose efficient organization was responsible for letting the impression that all was smooth.

To all participants our final thanks.

WINFRIED BRUNS   ARON SIMIS

List of Participants

J. F. Andrade (Salvador, BRAZIL)
V. Bayer (Vitória, BRAZIL)
P. Brumatti (Campinas, BRAZIL)
W. Bruns (Vechta, WEST GERMANY)
A. J. Engler (Campinas, BRAZIL)
M. Fléxor (Paris, FRANCE)
F. Q. Gouvêa (São Paulo, BRAZIL)
J. Herzog (Essen, WEST GERMANY)
Y. Lequain (Rio de Janeiro, BRAZIL)

E. Marcos (São Paulo, BRAZIL)
M. Miller (Columbia, S.C., USA)
L. Robbiano (Genova, ITALY)
A. Simis (Salvador, BRAZIL)
B. Ulrich (East Lansing, USA)
I. Vainsencher (Recife, BRAZIL)
W. V. Vasconcelos (New Brunswick, USA)
L. Washington (Washington, USA)

Contents

WINFRIED BRUNS

    Straightening laws on modules and their symmetric algebras 1

MARIA PIA CAVALIERE, MARIA EVELINA ROSSI AND GIUSEPPE VALLA

    On short graded algebras 21

JÜRGEN HERZOG

    A Homological approach to symbolic powers 32

CRAIG HUNEKE AND BERND ULRICH

    Generic residual intersections 47

LORENZO ROBBIANO AND MOSS SWEEDLER

    Subalgebra bases 61

PETER SCHENZEL

    Flatness and ideal-transforms of finite type 88

ARON SIMIS

    Topics in Rees algebras of special ideals 98

WOLMER V. VASCONCELOS

    Symmetric algebras 115

Straightening Laws on Modules and Their Symmetric Algebras

WINFRIED BRUNS*

Several modules M over algebras with straightening law A have a structure which is similar to the structure of A itself: M has a system of generators endowed with a natural partial order, a standard basis over the ring B of coefficients, and the multiplication $A \times M \to A$ satisfies a "straightening law". We call them *modules with straightening law*, briefly MSLs.

In section 1 we recall the notion of an algebra with straightening law together with those examples which will be important in the sequel. Section 2 contains the basic results on MSLs, whereas section 3 is devoted to examples: (i) powers of certain ideals and residue class rings with respect to them, (ii) "generic" modules defined by generic, alternating or symmetric matrices of indeterminates, (iii) certain modules related to differentials and derivations of determinantal rings. The essential homological invariant of a module is its depth. We discuss how to compute the depth of an MSL in section 4. The main tool are filtrations related to the MSL structure.

The last section contains a natural strengthening of the MSL axioms which under certain circumstances leads to a straightening law on the symmetric algebra. The main examples of such modules are the "generic" modules defined by generic and alternating matrices.

The notion of an MSL was introduced by the author in [Br.3] and discussed extensively during the workshop. The main differences of this survey to [Br.3] are the more detailed study of examples and the treatment of the depth of MSLs which is almost entirely missing in [Br.3]

1. Algebras with Straightening Laws

An algebra with straightening law is defined over a ring B of coefficients. In order to avoid problems of secondary importance in the following sections we will assume throughout that B is a noetherian ring.

Definition. Let A be a B-algebra and $\Pi \subset A$ a finite subset with partial order $\leq$. A is an *algebra with straightening law on Π (over B)* if the following conditions are satisfied:

(ASL-0) $A = \bigoplus_{i \geq 0} A_i$ is a graded B-algebra such that $A_0 = B$, Π consists of homogeneous elements of positive degree and generates A as a B-algebra.

(ASL-1) The products $\xi_1 \cdots \xi_m$, $m \geq 0$, $\xi_1 \leq \cdots \leq \xi_m$ are a free basis of A as a B-module. They are called *standard monomials*.

(ASL-2) (*Straightening law*) For all incomparable $\xi, \upsilon \in \Pi$ the product $\xi\upsilon$ has a representation

$$\xi\upsilon = \sum a_\mu \mu, \qquad a_\mu \in B, a_\mu \neq 0, \quad \mu \quad \text{standard monomial},$$

---

*Partially supported by DFG and GMD

satisfying the following condition: every μ contains a factor $\zeta \in \Pi$ such that $\zeta \leq \xi, \zeta \leq \upsilon$. (It is of course allowed that $\xi\upsilon = 0$, the sum $\sum a_\mu \mu$ being empty.)

The theory of ASLs has been developed in [Ei] and [DEP.2]; the treatment in [BV.1] also satisfies our needs. In [Ei] and [BV.1] B-algebras satisfying the axioms above are called graded ASLs, whereas in [DEP.2] they figure as graded ordinal Hodge algebras.

In terms of generators and relations an ASL is defined by its poset and the straightening law:

(1.1) Proposition. *Let A be an ASL on Π. Then the kernel of the natural epimorphism*

$$B[T_\pi : \pi \in \Pi] \longrightarrow A, \qquad T_\pi \longrightarrow \pi,$$

is generated by the relations required in (ASL-2), *i.e. the elements*

$$T_\xi T_\upsilon - \sum a_\mu T_\mu, \qquad T_\mu = T_{\xi_1} \cdots T_{\xi_m} \quad if \quad \mu = \xi_1 \cdots \xi_m.$$

See [DEP.2, 1.1] or [BV.1, (4.2)].

(1.2) Proposition. *Let A be an ASL on Π, and $\Psi \subset \Pi$ an ideal, i.e. $\psi \in \Psi, \phi \leq \psi$ implies $\phi \in \Psi$. Then the ideal $A\Psi$ is generated as a B-module by all the standard monomials containing a factor $\psi \in \Psi$, and $A/A\Psi$ is an ASL on $\Pi \setminus \Psi$ ($\Pi \setminus \Psi$ being embedded into $A/A\Psi$ in a natural way.)*

This is obvious, but nevertheless extremely important. First several proofs by induction on $|\Pi|$, say, can be based on (1.2), secondly the ASL structure of many important examples is established this way.

(1.3) Examples. (a) Let X be an $m \times n$ matrix of indeterminates over B, and $I_{r+1}(X)$ denote the ideal generated by the $r + 1$-minors (i.e. the determinants of the $r + 1 \times r + 1$ submatrices) of X. For the investigation of the ideals $I_{r+1}(X)$ and the residue class rings $A = B[X]/I_{r+1}(X)$ one makes $B[X]$ an ASL on the set $\Delta(X)$ of all minors of X. Denote by $[a_1, \ldots, a_t | b_1, \ldots, b_t]$ the minor with row indices $a_1, \ldots, a_t$ and column indices $b_1, \ldots, b_t$. The partial order on $\Delta(X)$ is given by

$$[a_1, \ldots, a_u | b_1, \ldots, b_u] \leq [c_1, \ldots, c_v | d_1, \ldots, d_v] \qquad \Longleftrightarrow$$
$$u \geq v \quad \text{and} \quad a_i \leq c_i, \ b_i \leq d_i, \ i = 1, \ldots, v.$$

Then $B[X]$ is an ASL on $\Delta(X)$; cf. [BV.1], Section 4 for a complete proof. Obviously $I_{r+1}(X)$ is generated by an ideal in the poset $\Delta(X)$, so A is an ASL on the poset $\Delta_r(X)$ consisting of all the i-minors, $i \leq r$.

(b) Another example needed below is given by "pfaffian" rings. Let $X_{ij}, 1 \leq i < j \leq n$, be a family of indeterminates over B, $X_{ji} = -X_{ij}, X_{ii} = 0$. The pfaffian of the alternating matrix $(X_{i_u i_v} : 1 \leq u, v \leq t)$, t even, is denoted by $[i_1, \ldots, i_t]$. The polynomial ring $B[X]$ is an ASL on the set $\Phi(X)$ of the pfaffians $[i_1, \ldots, i_t], i_1 < \cdots < i_t, t \leq n$. The pfaffians are partially ordered in the same way as the minors in (b). The residue class ring $A = B[X]/\operatorname{Pf}_{r+2}(X)$, $\operatorname{Pf}_{r+2}(X)$ being generated by the $(r + 2)$-pfaffians, inherits its ASL structure from $B[X]$ according to (1.2). The poset underlying A is denoted $\Phi_r(X)$. Note that the rings A are Gorenstein rings over a Gorenstein B—in fact factorial over a factorial B, cf. [Av.1], [KL].

(c) A non-example: If X is a symmetric $n \times n$ matrix of indeterminates, then $B[X]$ can *not* be made an ASL on $\Delta(X)$ in a natural way. Nevertheless there is a standard monomial theory for this ring based on the concept of a *doset*, cf. [DEP.2]. Many results which can be derived from this theory were originally proved by Kutz [Ku] using the method of principal radical systems. —

For an element $\xi \in \Pi$ we define its rank by

$$\text{rk}\, \xi = k \quad \Longleftrightarrow \quad \text{there is a chain } \xi = \xi_k > \xi_{k-1} > \cdots > \xi_1, \ \xi_i \in \Pi,$$
$$\text{and no such chain of greater length exists.}$$

For a subset $\Omega \subset \Pi$ let

$$\text{rk}\, \Omega = \max\{\text{rk}\, \xi : \xi \in \Omega\}.$$

The preceding definition differs from the one in [Ei] and [DEP.2] which gives a result smaller by 1. In order to reconcile the two definitions the reader should imagine an element $-\infty$ added to Π, vaguely representing $0 \in A$.

(1.4) Proposition. *Let A be an ASL on Π. Then*

$$\dim A = \dim B + \text{rk}\, \Pi \quad and \quad \text{ht}\, A\Pi = \text{rk}\, \Pi.$$

Here of course $\dim A$ denotes the Krull dimension of A and $\text{ht}\, A\Pi$ the height of the ideal $A\Pi$. A quick proof of (1.4) may be found in [BV.1, (5.10)].

2. Straightening Laws on Modules

It occurs frequently that a module M over an ASL A has a structure closely related to that of A: the generators of M are partially ordered, a distinguished set of "standard elements" forms a B-basis of M, and the multiplication $A \times M \to A$ satisfies a straightening law similar to the straightening law in A itself. In this section we introduce the notion of a module with straightening law whereas the next section contains a list of examples.

Definition. Let A be an ASL over B on Π. An A-module M is called a *module with straightening law* (MSL) on the finite poset $\mathcal{X} \subset M$ if the following conditions are satisfied:
(MSL-1) For every $x \in \mathcal{X}$ there exists an ideal $\mathcal{I}(x) \subset \Pi$ such that the elements

$$\xi_1 \cdots \xi_n x, \qquad x \in \mathcal{X}, \quad \xi_1 \notin \mathcal{I}(x), \quad \xi_1 \leq \cdots \leq \xi_n, \quad n \geq 0,$$

constitute a B-basis of M. These elements are called *standard elements*.
(MSL-2) For every $x \in \mathcal{X}$ and $\xi \in \mathcal{I}(x)$ one has

$$\xi x \in \sum_{y < x} Ay.$$

It follows immediately by induction on the rank of x that the element ξx as in (MSL-2) has a standard representation

$$\xi x = \sum_{y < x} \left(\sum b_{\xi x \mu y} \mu \right) y, \qquad b_{\xi x \mu y} \in B, \ b_{\xi x \mu y} \neq 0,$$

in which each μy is a standard element.

(2.1) Remarks. (a) Suppose M is an MSL, and $\mathcal{T} \subset \mathcal{X}$ an ideal. Then the submodule of M generated by $\mathcal{T}$ is an MSL, too. This fact allows one to prove theorems on MSLs by noetherian induction on the set of ideals of $\mathcal{X}$.

(b) It would have been enough to require that the standard elements are linearly independent. If just (MSL-2) is satisfied then the induction principle in (a) proves that M is generated as a B-module by the standard elements. —

The following proposition helps to detect MSLs:

(2.2) Proposition. *Let M, M_1, M_2 be modules over an ASL A, connected by an exact sequence*

$$0 \longrightarrow M_1 \longrightarrow M \longrightarrow M_2 \longrightarrow 0.$$

Let M_1 and M_2 be MSLs on $\mathcal{X}_1$ and $\mathcal{X}_2$, and choose a splitting f of the epimorphism $M \to M_2$ over B. Then M is an MSL on $\mathcal{X} = \mathcal{X}_1 \cup f(\mathcal{X}_2)$ ordered by $x_1 < f(x_2)$ for all $x_1 \in \mathcal{X}_1$, $x_2 \in \mathcal{X}_2$, and the given partial orders on $\mathcal{X}_1$ and the copy $f(\mathcal{X}_2)$ of $\mathcal{X}_2$. Moreover one chooses $\mathcal{I}(x)$, $x \in \mathcal{X}_1$, as in M_1 and $\mathcal{I}(f(x)) = \mathcal{I}(x)$ for all $x \in \mathcal{X}_2$.

The proof is straightforward and can be left to the reader.

In terms of generators and relations an ASL is defined by its generating poset and its straightening relations, cf. (1.1). This holds similarly for MSLs:

(2.3) Proposition. *Let A be an ASL on Π over B, and M an MSL on $\mathcal{X}$ over A. Let e_x, $x \in \mathcal{X}$, denote the elements of the canonical basis of the free module $A^{\mathcal{X}}$. Then the kernel $K_{\mathcal{X}}$ of the natural epimorphism*

$$A^{\mathcal{X}} \longrightarrow M, \qquad e_x \longrightarrow x,$$

is generated by the relations required for (MSL-2):

$$\rho_{\xi x} = \xi e_x - \sum_{y < x} a_{\xi xy} e_y, \qquad x \in \mathcal{X}, \; \xi \in \mathcal{I}(x).$$

PROOF: We use the induction principle indicated in (2.1), (a). Let $\tilde{x} \in \mathcal{X}$ be a maximal element. Then $\mathcal{T} = \mathcal{X} \setminus \{\tilde{x}\}$ is an ideal. By induction $A\mathcal{T}$ is defined by the relations $\rho_{\xi x}$, $x \in \mathcal{T}, \xi \in \mathcal{I}(x)$. Furthermore (MSL-1) and (MSL-2) imply

$$(1) \qquad\qquad M/A\mathcal{T} \cong A/A\mathcal{I}(\tilde{x})$$

If $a_{\tilde{x}}\tilde{x} - \sum_{y \in \mathcal{T}} a_y y = 0$, one has $a_{\tilde{x}} \in A\mathcal{I}(\tilde{x})$ and subtracting a linear combination of the elements $\rho_{\xi \tilde{x}}$ from $a_{\tilde{x}} e_{\tilde{x}} - \sum_{y \in \mathcal{T}} a_y e_y$ one obtains a relation of the elements $y \in \mathcal{T}$ as desired. —

The kernel of the epimorphism $A^{\mathcal{X}} \to M$ is again an MSL:

(2.4) Proposition. *With the notations and hypotheses of (2.3) the kernel $K_{\mathcal{X}}$ of the epimorphism $A^{\mathcal{X}} \to M$ is an MSL if we let*

$$\mathcal{I}(\rho_{\xi x}) = \{\pi \in \Pi : \pi \not\geq \xi\}$$

and

$$\rho_{\xi x} \leq \rho_{\upsilon y} \quad\Longleftrightarrow\quad x < y \;\; or \;\; x = y, \; \xi \leq \upsilon.$$

PROOF: Choose $\tilde{x}$ and $\mathcal{T}$ as in the proof of (2.3). By virtue of (2.3) the projection $A^{\mathcal{X}} \to Ae_{\tilde{x}}$ with kernel $A^{\mathcal{T}}$ induces an exact sequence

$$0 \longrightarrow K_{\mathcal{T}} \longrightarrow K_{\mathcal{X}} \longrightarrow A\mathcal{I}(\tilde{x}) \longrightarrow 0.$$

Now (2.2) and induction finish the argument. —

If a module M is given in terms of generators and relations, it is in general more difficult to establish (MSL-1) than (MSL-2). For (MSL-2) one "only" has to show that elements $\rho_{\xi x}$ as in the proof of (2.3) can be obtained as linear combinations of the given relations. In this connection the following proposition may be useful: it is enough that the module generated by the $\rho_{\xi x}$ satisfies (MSL-2) again.

(2.5) Proposition. *Let the data $M, \mathcal{X}, \mathcal{I}(x), x \in \mathcal{X}$, be given as in the definition, and suppose that (MSL-2) is satisfied. Suppose that the kernel $K_{\mathcal{X}}$ of the natural epimorphism $A^{\mathcal{X}} \to M$ is generated by the elements $\rho_{\xi x} \in A^{\mathcal{X}}$ representing the relations in (MSL-2). Order the $\rho_{\xi x}$ and choose $\mathcal{I}(\rho_{\xi x})$ as in (2.4). If $K_{\mathcal{X}}$ satisfies (MSL-2) again, M is an MSL.*

PROOF: Let $\tilde{x} \in \mathcal{X}$ be a maximal element, $\mathcal{T} = \mathcal{X} \setminus \{\tilde{x}\}$. We consider the induced epimorphism

$$A^{\mathcal{T}} \longrightarrow A\mathcal{T}$$

with kernel $K_{\mathcal{T}}$. One has $K_{\mathcal{T}} = K_{\mathcal{X}} \cap A^{\mathcal{T}}$. Since the $\rho_{\xi x}$ satisfy (MSL-2), every element in $K_{\mathcal{X}}$ can be written as a B-linear combination of standard elements, and only the $\rho_{\xi \tilde{x}}$ have a nonzero coefficient with respect to $e_{\tilde{x}}$. The projection onto the component $Ae_{\tilde{x}}$ with kernel $A^{\mathcal{T}}$ shows that $K_{\mathcal{T}}$ is generated by the $\rho_{\xi x}, x \in \mathcal{T}$. Now one can argue inductively, and the split-exact sequence

$$0 \longrightarrow A\mathcal{T} \longrightarrow M \longrightarrow M/A\mathcal{T} \cong A/A\mathcal{I}(\tilde{x}) \longrightarrow 0$$

of B-modules finishes the proof. —

Modules with a straightening law have a distinguished filtration with cyclic quotients; by the usual induction this follows immediately from the isomorphism (1) above:

(2.6) Proposition. *Let M be an MSL on $\mathcal{X}$ over A. Then M has a filtration $0 = M_0 \subset M_1 \subset \cdots \subset M_n = M$ such that each quotient M_{i+1}/M_i is isomorphic with one of the residue class rings $A/A\mathcal{I}(x), x \in \mathcal{X}$, and conversely each such residue class ring appears as a quotient in the filtration.*

It is obvious that an A-module with a filtration as in (2.6) is an MSL. It would however not be adequate to replace (MSL-1) and (MSL-2) by the condition that M has such a filtration since (MSL-1) and (MSL-2) carry more information and lend themselves to natural strengthenings, see section 5.

In section 4 we will base a depth bound for MSLs on (2.6). Further consequences concern the annihilator, the localizations with respect to prime ideals $P \in \text{Ass}\, A$, and the rank of an MSL.

(2.7) Proposition. *Let M be an MSL on $\mathcal{X}$ over A, and*

$$J = A(\bigcap_{x \in \mathcal{X}} \mathcal{I}(x)).$$

Then

$$J \supset \operatorname{Ann} M \supset J^n, \qquad n = \operatorname{rk} \mathcal{X}.$$

PROOF: Note that $A(\bigcap \mathcal{I}(x)) = \bigcap A\mathcal{I}(x)$ (as a consequence of (1.2)). Since $\operatorname{Ann} M$ annihilates every subquotient of M, the inclusion $\operatorname{Ann} M \subset J$ follows from (2.6). Furthermore (MSL-2) implies inductively that

$$J^i M \subset \sum_{\operatorname{rk} x \leq \operatorname{rk} \Pi - i} Ax$$

for all i, in particular $J^n M = 0$. —

(2.8) Proposition. *Let M be an MSL on $\mathcal{X}$ over A, and $P \in \operatorname{Ass} A$.*
(a) *Then $\{\pi \in \Pi : \pi \not\subset P\}$ has a single minimal element σ, and σ is also a minimal element of Π.*
(b) *Let $\mathcal{Y} = \{x \in \mathcal{X} : \sigma \notin \mathcal{I}(x)\}$. Then $\mathcal{Y}$ is a basis of the free A_P-module M_P. Furthermore $(K_{\mathcal{X}})_P$ is generated by the elements $\varrho_{\sigma x}$, $x \notin \mathcal{Y}$.*

PROOF: (a) If π_1, π_2, $\pi_1 \neq \pi_2$, are minimal elements of $\{\pi \in \Pi : \pi \not\subset P\}$, then, by (ASL-2), $\pi_1 \pi_2 \in P$. So there is a single minimal element σ. It has to be a single minimal element of Π, too, since otherwise P would contain all the minimal elements of Π whose sum, however, is not a zero-divisor in A ([BV.1, (5.11)]).

(b) Consider the exact sequence

$$0 \longrightarrow A\mathcal{T} \longrightarrow M \longrightarrow A/A\mathcal{I}(\tilde{x}) \longrightarrow 0$$

introduced in the proof of (2.3). If $\tilde{x} \notin \mathcal{Y}$, then $\tilde{x} \in A_P \mathcal{T}$ by the relation $\varrho_{\sigma \tilde{x}}$, and we are through by induction. If $\tilde{x} \in \mathcal{Y}$, then σ and all the elements of $\mathcal{I}(\tilde{x})$ are incomparable, so they are annihilated by σ (because of (ASL-2)). Consequently $(A/A\mathcal{I}(\tilde{x}))_P \cong A_P$, $\tilde{x}$ generates a free summand of M_P, and induction finishes the argument again. —

We say that a module M over A has rank r if $M \otimes L$ is free of rank r as an L-module, L denoting the total ring of fractions of A. Cf. [BV.1, 16.A] for the properties of this notion.

(2.9) Corollary. *Let M be an MSL on $\mathcal{X}$ over the ASL A on Π. Suppose that Π has a single minimal element π, a condition satisfied if A is a domain. Then*

$$\operatorname{rank} M = |\{x \in \mathcal{X} : \mathcal{I}(x) = \emptyset\}|.$$

3. Examples

In this section we list some of the examples of MSLs. The common patterns in their treatment in [BV.1], [BV.2], and [BST] were the author's main motivation in the creation of the concept of an MSL. We start with a very simple example:

(3.1) Example. A itself is an MSL if one takes $\mathcal{X} = \{1\}$, $\mathcal{I}(1) = \emptyset$. Another choice is $\mathcal{X} = \Pi \cup \{1\}$, $\mathcal{I}(\xi) = \{\pi \in \Pi : \pi \not\geq \xi\}$, $\mathcal{I}(1) = \Pi$, $1 > \pi$ for each $\pi \in \Pi$. The relations necessary for (MSL-2) are then given by the identities $\pi 1 = \pi$, the straightening relations

$$\xi \upsilon = \sum b_\mu \mu, \qquad \xi, \upsilon \text{ incomparable,}$$

and the Koszul relations

$$\xi \upsilon = \upsilon \xi, \qquad \xi < \upsilon.$$

By (2.1),(a) for every poset ideal $\Psi \subset \Pi$ the ideal $A\Psi$ is an MSL, too.

(3.2) MSLs derived from powers of ideals. (a) Suppose that Ψ as in (3.1) additionally satisfies the following condition: Whenever $\phi, \psi \in \Psi$ are incomparable, then every standard monomial μ in the standard representation $\phi\psi = \sum a_\mu \mu$, $a_\mu \neq 0$, contains at least two factors from Ψ. This condition appears in [Hu], [EH], and in [BV.1, Section 9] where the ideal $I = A\Psi$ is called *straightening-closed*. See [BST] for a detailed treatment of straightening-closed ideals. As a consequence of (b) below the powers I^n of $I = A\Psi$ are MSLs. Observe in particular that the condition above is satisfied if every μ a priori contains at most two factors and Ψ consists of the elements in Π of highest degree.

(b) In order to prove and to generalize the statements in (a) let us consider an MSL M on $\mathcal{X}$ and an ideal $\Psi \subset \Pi$ such that $I = A\Psi$ is straightening-closed and the following condition holds:

($*$) The standard monomials in the standard representation of a product ψx, $\psi \in \Psi$, $x \in \mathcal{X}$, all contain a factor from Ψ.

Then it is easy to see that IM is again an MSL on the set $\{\psi x : x \in \mathcal{X}, \psi \in \Psi \setminus \mathcal{I}(x)\}$ partially ordered by

$$\psi x \leq \phi y \quad \Longleftrightarrow \quad x < y \quad \text{or} \quad x = y, \ \psi \leq \phi,$$

if one takes

$$\mathcal{I}(\psi x) = \{\pi \in \Pi : \pi \not\geq \psi\}.$$

Furthermore ($*$) holds again. Thus $I^n M$ is an MSL for all $n \geq 1$, and in particular one obtains (b) from the special case $M = A$.

The residue class module M/IM also carries the structure of an MSL on the set $\overline{\mathcal{X}}$ of residues of $\mathcal{X}$ if we let

$$\mathcal{I}(\overline{x}) = \mathcal{I}(x) \cup \Psi.$$

Combining the previous arguments we get that $I^n M/I^{n+1}M$ is an MSL for all $n \geq 0$. Arguing by (2.2) one sees that all the quotients $I^n M/I^{n+k}M$ are MSLs.

In the situation just considered the associated graded ring $\mathrm{Gr}_I A$ is an ASL on the set Π^* of leading forms (ordered in the same way as Π), cf. [BST] or [BV.1,(9.8)], and obviously $\mathrm{Gr}_I M$ is an MSL on $\mathcal{X}^*$.

(c) If an ideal $I = A\Psi$ is not straightening-closed, one cannot make the associated graded ring an ASL in a natural way. Under certain circumstances there is however a "canonical" substitute, the *symbolic associated graded ring*

$$\mathrm{Gr}_I^{()}(A) = \bigoplus_{i=0}^{\infty} I^{(i)}/I^{(i+1)}.$$

Suppose that every standard monomial in a straightening relation of A contains at most two factors and that Ψ consists of all the elements of Π whose degree is at least d, d fixed. Furthermore put

$$\gamma(\pi) = \begin{cases} 0 & \text{if } \deg \pi < d, \\ \deg \pi - d + 1 & \text{else,} \end{cases} \quad \text{and} \quad \gamma(\pi_1 \dots \pi_m) = \sum \gamma(\pi_i)$$

for an element $\pi \in \Pi$ and a standard monomial $\pi_1 \dots \pi_m$ (deg denotes the degree in the graded ring A). Then it is not difficult to show that the B-submodule J_i generated by the standard monomials μ such that $\gamma(\mu) \geq i$ is an ideal of A and that $\bigoplus J_i/J_{i+1}$ is (a well-defined B-algebra and) an ASL over B on the poset given by the leading forms of the elements of Π cf. [DEP.2, Section 10]. Therefore J_i and J_i/J_{i+1} have standard B-bases and one easily establishes that they are MSLs.

For $B[X]$, B a domain, X a generic matrix of indeterminates or an alternating matrix of indeterminates, J_i indeed is the i-th symbolic power of the ideal I generated by all minors or pfaffians resp. of size d, [BV.1, 10.A] or [AD]. Consequently $\mathrm{Gr}_I^{()}(A)$ is an ASL, and $I^{(i)}$, $I^{(i)}/I^{(i+1)}$ are MSLs for all i.

(3.3) MSLs derived from generic maps. (a) Let $A = B[X]/I_{r+1}(X)$ as in (1.3), (a), $0 \leq r \leq \min(m,n)$ (so $A = B[X]$ is included). The matrix x over A whose entries are the residue classes of the indeterminates defines a map $A^m \to A^n$, also denoted by x. The modules $\mathrm{Im}\, x$ and $\mathrm{Coker}\, x$ have been investigated in [Br.1]. A simplified treatment has been given in [BV.1, Section 13], from where we draw some of the arguments below. Let $d_1, \ldots, d_m$ and $e_1, \ldots, e_n$ denote the canonical bases of A^m and A^n. Then we order the system $\bar{e}_1, \ldots, \bar{e}_n$ of generators of $M = \mathrm{Coker}\, x$ linearly by

$$\bar{e}_1 > \cdots > \bar{e}_n.$$

Furthermore we put

$$\mathcal{I}(\bar{e}_i) = \begin{cases} \{\delta \in \Delta_r(X) : \delta \not\geq [1,\ldots,r|1,\ldots,\hat{i},\ldots,r+1]\} & \text{for } i \leq r, \\ \emptyset & \text{else,} \end{cases}$$

if $r < n$, and in the case in which $r = n$

$$\mathcal{I}(\bar{e}_i) = \{\delta \in \Delta_r(X) : \delta \not\geq [1,\ldots,r-1|1,\ldots,\hat{i},\ldots,r]\}.$$

(where $\hat{i}$ denotes that i is to be omitted). We claim: M is an MSL with respect to these data.

Suppose that $\delta \in \mathcal{I}(\bar{e}_i)$. Then

$$\delta = [a_1,\ldots,a_s|1,\ldots,i,b_{i+1},\ldots,b_s], \qquad s \leq r.$$

The element

$$\sum_{j=1}^{s} (-1)^{j+i} [a_1,\ldots,\hat{a}_j,\ldots,a_s|1,\ldots,i-1,b_{i+1},\ldots,b_s] x(d_{a_j})$$

of $\mathrm{Im}\, x$ is a suitable relation for (MSL-2):

(1) $$\delta \bar{e}_i = \sum_{k=i+1}^{n} \pm [a_1,\ldots,a_s|1,\ldots,i-1,k,b_{i+1},\ldots,b_s] \bar{e}_k.$$

Rearranging the column indices $1,\ldots,i-1,k,b_{i+1},\ldots,b_s$ in ascending order one makes (1) the standard representation of $\delta \bar{e}_i$, and observes the following fact recorded for later purpose:

(2) $\delta \notin \mathcal{I}(\bar{e}_k)$ for all $k \geq i+1$ such that $[a_1,\ldots,a_s|1,\ldots,i-1,k,b_{i+1},\ldots,b_s] \neq 0$.

In order to prove the linear independence of the standard elements one may assume that $r < n$ since $I_n(X)$ annihilates M. Let

$$\widetilde{M} = \sum_{i=r+1}^{n} A\bar{e}_i, \quad \Psi = \{\delta \in \Delta_r(X): \delta \not\geq [1,\ldots,r|1,\ldots,r-1,r+1]\} \quad \text{and} \quad I = A\Psi.$$

We claim:
(i) $\widetilde{M}$ is a free A-module.
(ii) $M/\widetilde{M}$ is (over A/I) isomorphic to the ideal generated by the minors $[1,\ldots,r|1,\ldots,\widehat{i},$ $\ldots,r+1]$, $1 \leq i \leq r$, in A/I.
In fact, the minors just specified form a linearly ordered ideal in the poset $\Delta_r(X) \setminus \Psi$ underlying the ASL A/I, and the linear independence of the standard elements follows immediately from (i) and (ii).

Statement (i) simply holds since rank $x = r$, and the r-minor in the left upper corner of x, being the minimal element of $\Delta_r(X)$, is not a zero-divisor in A. For (ii) one applies (2.3) to show that $M/\widetilde{M}$ and the ideal in (ii) have the same representation given by the matrix

$$\begin{pmatrix} x_{11} & \cdots & x_{1r} \\ \vdots & & \vdots \\ x_{m1} & \cdots & x_{mr} \end{pmatrix},$$

the entries taken in A/I: The assignment $\bar{e}_i \to (-1)^{i+1}[1,\ldots,r|1,\ldots,\widehat{i},\ldots,r+1]$ induces the isomorphism. The computations needed for the application of (2.5) are covered by (1).

By similar arguments one can show that Im x is also an MSL, see [BV.1, proof of (13.6)] where a filtration argument is given which shows the linear independence of the standard elements. Such a filtration argument could also have been applied to prove (MSL-1) for M, cf. (c) below.

(b) Another example is furnished by the modules defined by generic alternating maps. Recalling the notations of (1.3), (b) we let $A = B[X]/\operatorname{Pf}_{r+2}(X)$ and M be the cokernel of the linear map

$$x: F \longrightarrow F^*, \qquad F = A^n.$$

In complete analogy with the preceding example M is an MSL on $\{\bar{e}_1,\ldots,\bar{e}_n\}$, the canonical basis of F^*, $\bar{e}_1 > \cdots > \bar{e}_n$, if one puts

$$\mathcal{I}(\bar{e}_i) = \begin{cases} \{\pi \in \Phi_r(X) : \pi \not\geq [1,\ldots,\widehat{i},\ldots,r+1]\} & \text{for} \quad i \leq r, \\ \emptyset & \text{else,} \end{cases}$$

if $r < n$, and in the case in which $r = n$

$$\mathcal{I}(\bar{e}_i) = \begin{cases} \{\pi \in \Phi(X) : \pi \not\geq [1,\ldots,\widehat{i},\ldots,r-1]\} & \text{for} \quad i \leq n-1, \\ \{[1,\ldots,n]\} & \text{for} \quad i = n. \end{cases}$$

The straightening law (1) is replaced by the equation

$$(1') \qquad \pi\bar{e}_i = \sum_{k=i+1}^{n} \pm[1,\ldots,i-1,k,b_{i+1},\ldots,b_s]\bar{e}_k,$$

obtained from Laplace type expansion of pfaffians as (1) has been derived from Laplace expansion of minors. Observe that the analogue $(2')$ of (2) is satisfied. The linear independence of the standard elements is proved in entire analogy with (d). With $\widetilde{M} = \sum_{i=r+1}^{n} A\bar{e}_i$ and $I = A[1,\dots,r]$ one has in the essential case $r < n$:

(i') $\widetilde{M}$ is a free A-module.

(ii') $M/\widetilde{M}$ is (over A/I) isomorphic to the ideal generated by the pfaffians $[1,\dots,\hat{i},\dots, r+1]$, $1 \le i \le r$, in A/I.

A notable special case is n odd, $r = n - 1$. In this case Coker $x \cong \mathrm{Pf}_r(X)$ is an ideal of grade 2 and projective dimension 2 [BE] and generated by a linearly ordered poset ideal in $\Phi(X)$.

(c) The two previous examples suggest to discuss the case of a symmetric matrix of indeterminates as in (1.3),(c), too. As mentioned there, the ring $A = B[X]/I_{r+1}(X)$ is not an ASL. Nevertheless the cokernel M of the map $x\colon F \to F^*$, $F = A^n$, has the same structure relative to A as the modules in the two previous examples. With respect to what is known about the rings A, it is easier to work with slightly different arguments which could have been applied in (a) and (b), too, and were in fact applied in [BV.1] to the modules of (a).

Taking analogous notations as in (b), we put $M_i = \sum_{j=i+1}^{n} A\bar{e}_j$, $\bar{e}_j$ denoting the residue class in M of the j-th canonical basis element of F^*. One has a filtration

$$M = M_0 \supset M_1 \supset \cdots \supset M_r.$$

We claim:

(i) M_r is a free A-module.

(ii) The annihilator J_i of M/M_i is the ideal generated by the i-minors of the first i columns of x.

(iii) The generator $\bar{e}_i$ of M_{i-1}/M_i is linearly independent over A/J_i.

Claim (i) is clear: rank $x = r$, and the first r columns are linearly independent, hence rank $M/M_r = 0 = \mathrm{rank}\, M - (n - r)$—none of the r-minors of x is a zero-divisor of A by the results of Kutz [Ku]. (This may not be found explicitly in [Ku] for arbitrary B, it is however enough to have it over a field B, cf. [BV.1, (3.15)]). Since M/M_i is represented by the matrix $(x \,|\, i)$ consisting of the first i columns of x, Ann $M/M_i \supset J_i$. On the other hand the first $i - 1$ columns of $(x \,|\, i)$ are linearly independent over A/J_i (again by [Ku]), and by the same argument as used for (i) one concludes (iii) and (ii).

Altogether M has a filtration by cyclic modules whose structure can be considered well-understood because of the results of [Ku] or the standard basis arguments based on the notion of a doset [DEP.2]. In particular M is a free B-module. Taking into account the remark below (2.6) one sees that one could call M an MSL relative to A. Of course the modules in (a) and (b) have an analogous filtration as follows from (2.6). —

(3.4) MSLs related to differentials and derivations. Let $A = B[X]/I_{r+1}(X)$. The module $\Omega = \Omega_{A/B}$ of Kähler differentials of A and its dual Ω^*, the module of derivations, have been investigated in [Ve.1], [Ve.2], and [BV.1]. A crucial point in the investigation of Ω is a filtration which stems from an MSL structure on the first syzygy of Ω. In fact, with $I = I_{r+1}(X)$, one has an exact sequence

$$0 \longrightarrow I/I^{(2)} \longrightarrow \Omega_{B[X]/B} \otimes A \longrightarrow \Omega \longrightarrow 0,$$

and it has been observed in (3.2),(c) that $I/I^{(2)}$ is an MSL.

The surjection $\Omega_{B[X]/B} \otimes A \longrightarrow \Omega$ induces an embedding $\Omega^* \longrightarrow (\Omega_{B[X]/B} \otimes A)^*$ whose cokernel is denoted N in [BV.1, Section 15]. It follows immediately from the filtration described in [BV.1, (15.3)] that N is an MSL. (It would take too much space to describe this filtration in such a detail that would save the reader to look up [BV.1].)

4. The depth of an MSL

As usual let A be an ASL over B on Π. For any A-module M we denote the length of a maximal M-sequence in $A\Pi$ by depth M. An MSL M over A is free as a B-module, in particular flat. Let P be a prime ideal of A, $P \supset A\Pi$, and put $Q = P \cap B$, $\kappa(Q) = B_Q/QB_Q$. By [Ma, (21.B)] one has

$$\text{depth } M_P = \text{depth } B_Q + \text{depth}(M \otimes \kappa(Q))_P.$$

Since all the prime ideals Q of B appear in the form $P \cap B$, it turns out that

$$\text{depth } M = \min_P \text{depth}(M \otimes \kappa(Q))_P, \qquad Q = P \cap B.$$

One sees easily that $M \otimes \kappa(Q)$ is an MSL over $A \otimes \kappa(Q)$, an ASL over $\kappa(Q)$. Therefore eventually

$$\text{depth } M = \min_Q \text{depth } M \otimes \kappa(Q).$$

This means: In computing depth M only the case in which B is a field is essential, and if the result does not depend on the particular field (as will be the case below) it holds automatically for arbitrary B. (Another possibility very often is the reduction to the case $B = \mathbf{Z}$ in order to apply results on generic perfection, cf. [BV.1], [BV.2].)

Every MSL has a natural filtration by (2.6). Applying the standard result on the behaviour of depth along short exact sequences one therefore obtains:

(4.1) Proposition. *Let M be an MSL on $\mathcal{X}$ over A. Then*

$$\text{depth } M \geq \min\{\text{depth } A/A\mathcal{I}(x) \colon x \in \mathcal{X}\}.$$

We specialize to ASLs over wonderful posets (cf. [Ei], [DEP.2], or [BV.1] for this notion and the properties of ASLs over wonderful posets).

(4.2) Corollary. *Let A be an ASL on the wonderful poset Π. If M is an MSL on $\mathcal{X}$ over A, then*

$$\text{depth } M \geq \min\{\text{rk } \Pi - \text{rk } \mathcal{I}(x) \colon x \in \mathcal{X}\}.$$

Since M may be the direct sum of the quotients in its natural filtration there is no way to give a better bound for depth M in general. Even when (4.2) does not give the best possible result it may be useful as a "bootstrap". While it is sometimes possible to find a coarser filtration which preserves more of the structure of M, there are also examples for which the exact computation of depth M requires completely different, additional arguments. We now discuss the examples in the same order as in the preceding section.

(4.3) MSLs derived from powers of ideals. As in (3.2) let $I = A\Psi$ be straightening-closed. Applying (4.2) to I^n and changing to A/I^n then, one obtains:

(a) *Suppose that* Π *is wonderful. Then* $\min_i \operatorname{depth} A/I^i \geq \operatorname{rk}\Pi - \operatorname{rk}\Psi$.

Elementary examples show that (a) is by no means sharp in general: Take $A = B[X]$, X a 2×2 matrix, I the ideal generated by the elements in its first column. Then obviously $\operatorname{depth} A/I^i = 2$ for all i, and (a) gives the lower bound 2 if one takes $\Pi = \{X_{11}, X_{21}, X_{12}, X_{22}\}$, its elements ordered in the sequence given. On the other hand, the choice $\Pi = \Delta(X)$ gives the lower bound 1 only since Ψ then consists of X_{11}, X_{21}, and $[1\ 2|1\ 2]$, hence $\operatorname{rk}\Psi = 3$. Under special hypotheses the bound given by (a) is sharp however:

(b) *Suppose, in addition, that* Ψ *consists of elements of highest degree within* Π *and that the standard monomials in the straightening relations of* A *have at most two factors. Then* $\min_i \operatorname{depth} A/I^i = \operatorname{rk}\Pi - \operatorname{rk}\Psi$.

This is [BST, (3.3.3)]. We sketch its proof: First one reduces the problem to the case of a field B as above. Then one shows that $\operatorname{Gr}_I A/\Pi \operatorname{Gr}_I A$ is isomorphic to the sub-ASL of A generated by the elements of Ψ. The latter obviously has dimension $\operatorname{rk}\Psi$. Thus one knows the analytic spread $\ell(I)$ and obtains $\min_i \operatorname{depth} A/I^i = \dim A - \ell(I) = \operatorname{rk}\Pi - \operatorname{rk}\Psi$ since $\operatorname{Gr}_I A$ is a Cohen-Macaulay ring.

Completely analogous arguments can be applied to derive the same result for the ideals discussed in (3.2),(c):

(c) *Suppose that the monomials in the straightening relations of* A *have at most two factors, and let* Ψ *be the ideal of* Π *generated by the elements of degree at least* d, d *fixed. Then, with the notations of* (3.3),(c) *one has:* $\min_i \operatorname{depth} A/J_i = \operatorname{rk}\Pi - \operatorname{rk}\Psi$.

See [BV.1, 10.B] for the case $A = B[X]$, $\Pi = \Delta(X)$, $J = I_d(X)$ in which, as mentioned in (3.3),(c) already, $J_i = I_d(X)^{(i)}$.

It would be interesting to find natural filtrations on the modules I^i (or A/I^i or I^i/I^{i+1}) and J_i in order to obtain a good lower bound for the depth of each individual power. This may be possible in special cases only. The instances for which we know $\operatorname{depth} R/I^i$ precisely for all n have been discussed in [BV.1, (9.27)]. Note that these results are based on free resolutions rather than filtrations.

(4.4) MSLs derived from generic maps. (a) Let first X be an $m \times n$ matrix of indeterminates, and $A = R_{r+1}(X)$. We consider the map $x\colon A^m \to A^n$ as in (3.3),(a) and its cokernel M. In determining $\operatorname{depth} M$ we assume rightaway that B is a field. Since $I_n(X)$ annihilates M the case $r = n$ is covered by the case $r = n - 1$; therefore one can restrict oneself to the case $r < n$. As shown in (3.3),(a) M fits into an exact sequence

$$0 \longrightarrow \widetilde{M} \longrightarrow M \longrightarrow J \longrightarrow 0$$

in which $\widetilde{M}$ is free over A and J is an ideal in A/I, I generated by the r-minors of the first r columns of x. It is not difficult to show via (1.4) that $\operatorname{depth} J = \operatorname{depth} A - 1$: A/I and $(A/I)/J$ are Cohen-Macaulay again, and the dimensions of A, A/I, and $(A/I)/J$ differ successively by 1, cf. [BV.1, proof of (13.4)]. This implies

$$\operatorname{depth} M \geq \operatorname{depth} A - 1.$$

It turns out that this inequality is an equation exactly when $m \geq n$, equivalently: $\mathrm{Ext}^1_R(M, \omega_A) = 0$ if and only if $m \geq n$. Fortunately the computations needed to prove this are not difficult—see [BV.1, 13.B] for the details.

(b) The case in which X is an alternating matrix and $A = B[X]/\mathrm{Pf}_{r+2}(X)$ is simpler such that we can give complete arguments relative to standard results on the rings A and ASLs in general.

There is one exceptional case: $n = r + 1$. As stated in (3.3),(b) already, one has $M \cong \mathrm{Pf}_r(X)$, whence M is an ideal of grade 3 and projective dimension 2 in this case. In particular depth $M =$ depth $A - 2$.

Similarly to (a) one can now restrict oneself to the case $r + 1 < n$. Using the exact sequence analogous to the one in (a) we get depth $M \geq$ depth $A - 1$: The defining ideal of $(A/I)/J$ as a residue class ring of A is generated by the pfaffians $\{\pi \in \Phi_r(X): \pi \not\geq [1,\ldots,r-1,r+2]\}$. Therefore $(A/I)/J$ is an ASL over a wonderful poset, cf. [BV.1, (5.10)]. Furthermore, computing the ranks of the underlying posets, one sees that the dimensions of A, A/I, and $(A/I)/J$ behave as in (a). (Note that in the exceptional case dealt with above $\dim A/I = \dim(A/I)/J + 2$.)

Since the matrix x is skew-symmetric, $M^* = \mathrm{Hom}_A(M, A) \cong \mathrm{Ker}\, x$, hence depth $M^* \geq \min(\text{depth}\, M + 2, \text{depth}\, A) = $ depth A. Furthermore A is a Gorenstein ring (over any Gorenstein B), cf. [KL]. Since M^* is a maximal Cohen-Macaulay module, its dual M^{**} is also a maximal Cohen-Macaulay module. Now it follows that M itself is a maximal Cohen-Macaulay module over A, since M is reflexive: The inequality depth $M \geq$ depth $A - 1$ carries over to all localizations of M and A. A well-known criterion for reflexivity (see [BV.1, 16.E] for example) therefore implies that it is enough to have M_P free over A_P for all prime ideals P of A such that depth $A_P \leq 2$. M_P is free if and only if P does not contain one of the r-pfaffians; the ideal generated by them in A has height $2(n - r) + 1 \geq 5$.

(c) The main arguments in (a) and (b) are first the isomorphism $M/\widetilde{M} \cong J$ together with precise information on depth J and secondly a duality argument. While the isomorphism could be established in the case of a symmetric matrix X as well and the duality argument will be used below, one lacks information on depth J. This forces us into a trickier line of proof which demonstrates the "bootstrap" function of a preliminary depth bound based on the filtration by cyclic modules as established in (3.3),(c). Again we assume that B is a field and that $r < n$.

(i) *If* $n \equiv r + 1$ (2), *then* depth $M =$ depth A.
(ii) *If* $n \not\equiv r + 1$ (2), *then* depth $M =$ depth $A - 1$.

Part (i) is almost as easy to prove as the same equation in (b). First we establish the depth bound based on the filtration by cyclic modules:

(iii) *For all* n *and all* r *one has*

$$\text{depth}\, M \geq \text{depth}\, A - r \geq \frac{1}{2}\,\text{depth}\, A.$$

In fact, by [Ku]

$$\text{depth}\, A \geq nr - r(r - 1)/2,$$

implying the second inequality. In (3.3),(c) we established that M has a filtration with quotients A and A/J_i, $i = 1,\ldots,r$. By [Ku] all these rings are Cohen-Macaulay, and $\dim A/J_i = \dim A + i - r - 1$. This proves the first inequality.

We now introduce a standard induction argument (which exists similarly under the conditions of (a) or (b) but was not necessary there). Take any prime ideal $Q \neq I_1(x)$ in A. Then there is (1) an element $x_{ii} \notin Q$ or (2) a 2-minor $x_{ii}x_{jj} - (x_{ij})^2 \notin Q$, by symmetry $x_{11} \notin Q$ or $x_{11}x_{22} - (x_{12})^2 \notin Q$. Over $B[X][X_{11}^{-1}]$ one performs elementary row and column transformations to obtain

$$\begin{pmatrix} X_{11} & 0 & \cdots & 0 \\ 0 & Y_{11} & \cdots & Y_{1,n-1} \\ \vdots & \vdots & \ddots & \vdots \\ 0 & Y_{1,n-1} & \cdots & Y_{n-1,n-1} \end{pmatrix},$$

$Y_{ij} = Y_{ji} = X_{i+1,j+1}X_{11} - X_{1,i+1}X_{1,j+1}$. It is easy to see that the elements Y_{ij}, $1 \leq i \leq j \leq n$, are algebraically independent over B and that $A[x_{11}^{-1}]$ is a Laurent polynomial extension of $B[Y]/I_r(Y)$. A similar argument works in case (2), now reducing both n and r by 2.

(iv) *There are families Y_{ij}, $1 \leq i \leq j \leq n-1$, and Z_{ij}, $1 \leq i \leq j \leq n-2$, of algebraically independent elements over B such that $A[x_{11}^{-1}]$ is a Laurent polynomial extension of $S_{r-1} = B[Y]/I_r(Y)$, and $A[(x_{11}x_{22}-x_{12}^2)^{-1}]$ is a Laurent polynomial extension of $T_{r-2} = B[Z]/I_{r-1}(Z)$. In both cases M is the extension of the modules defined by Y and Z resp.*

Now we can already prove (i) under whose hypotheses A is a Gorenstein ring. Let $P \subset A$ be the irrelevant maximal ideal. Arguing inductively via (iv) one may suppose that M_Q is a maximal Cohen-Macaulay module for all primes Q different from P. Let $D = \text{Coker } x^*$ be the Auslander-Bridger dual of M. Because x is symmetric, $D \cong M$. The assumptions so far imply that M_P is a d-th syzygy module, $d = \text{depth } M_P$, hence

$$\text{Ext}^i_{A_P}(M_P, A_P) = \text{Ext}^i_{A_P}(D_P, A_P) = 0 \qquad \text{for} \quad i = 1, \ldots, d,$$

(cf. [BV.1, 16.E] for example). On the other hand depth $M_P \geq d$ is equivalent to

$$\text{Ext}^i_{A_P}(M_P, A_P) = 0 \qquad \text{for} \quad i = \text{depth } A_P - d + 1, \ldots, \text{depth } A_P$$

by local duality. Hence $\text{Ext}^i_{A_P}(M_P, A_P) = 0$ for all $i > 0$, and M_P is a maximal Cohen-Macaulay module. This establishes (i).

Next we show that depth $M < $ depth A under the hypotheses of (ii). Again induction via (iv) can be applied to reduce to the case $r = 1$ first. Then $\text{Ext}^1_A(M, \omega_A) \neq 0$ is obvious since ω_A is generated by the entries of the first row (or column) of x, cf. [Go].

It remains to verify that depth $M \geq $ depth $A-1$ in (ii). Since depth $A/J_r = $ depth $A-1$, it is enough to show the following statements which hold for all n and all r:

(v) *As an (A/J_r)-module M/M_r is reflexive.*
(vi) *Its dual over A/J_r is isomorphic to J_{r-1}/J_r.*
(vii) *M/M_r is a maximal Cohen-Macaulay module over A/J_r.*
(In order to include the case $r = 1$: A 0-minor has the value 1.)

To simplify the notation write $\overline{A}$ for A/J_r and $\overline{M}$ for M/M_r. Let us first observe that (vii) holds in case $n \equiv r + 1$ (2) since, as has just been proved, M is a maximal Cohen-Macaulay module over A.

Next one notices that the case $r = 1$ is indeed trivial, M/M_1 being free of rank 1 over A/J_1. Suppose that $r > 1$ and proceed by induction. Then, via M and (iv), it follows that $\overline{M}_P$ is a maximal Cohen-Macaulay module over $\overline{A}_P$ for all $P \in \operatorname{Spec}\overline{A}$, $P \not\supset I_1(x)/J_r$.

For (v) it is enough to show that (1) $\overline{M}_P$ is free for all primes P such that depth $A_P \leq 1$, and (2) depth $\overline{M}_P \geq 2$ for the remaining ones. (1) is clear: grade $I_{r-1}(x \mid r)/J_r \geq 2$, and $\overline{M}_P$ is free if $P \not\supset I_{r-1}(x \mid r)/J_r$. In order to verify (2) one may now assume that $n \geq r+2$, $r > 1$, and $P \supset I_1(x)/J_r$. Then (iii) implies (2).

The dual of $\overline{M}$ is isomorphic to the kernel of the map $\overline{A}^r \rightarrow \overline{A}^n$ defined by the transpose y of $(x \mid r)$. Taking the determinantal relations of the rows of y, one sees that J_{r-1}/J_r is embedded in $\operatorname{Ker} y$ such that this embedding splits at all prime ideals not containing $I_{r-1}(x \mid r)/J_r$, in particular at all primes P such that depth $A_P \leq 1$. Since J_{r-1}/J_r is a maximal Cohen-Macaulay module over $\overline{A}$, (vi) follows easily.

It remains to prove (vii) for $n \not\equiv r + 1$ (2). In this case J_r is the canonical module of A, so $\overline{A} = A/J_r$ is a Gorenstein ring, cf. [HK, 6.13]. By (vi) the dual of $\overline{M}$ is Cohen-Macaulay, so is $\overline{M}$ by (v).

The results of (b) and (c) are also contained in [BV.2].

(4.5) MSLs related to differentials and derivations. We resume the hypotheses and notations of (3.4). One obtains a first depth bound for $I/I^{(2)}$ from (4.3),(c) above which is already quite good; it suffices to prove that Ω is reflexive. In order to get a precise result one has however to work with a coarser filtration, cf. [BV.1, Section 14].

A similar filtration yields that depth $N \geq$ depth $A - 2$, so depth $\Omega^* \geq$ depth $A - 1$ for all values of m, n, and r. While Ω^* cannot be a maximal Cohen-Macaulay module for a determinantal ring A if A is a non-regular Gorenstein ring, i.e. when $m = n$, $1 \leq r < \min(m,n)$, it has this property in all the other cases. Similar to (4.4),(a) this is shown by verifying that $\operatorname{Ext}_A^1(\Omega^*, \omega_A) = 0$. Unfortunately the details of this computation, for which we refer the reader to [BV.1, Section 15], are rather complicated.

5. Modules with a Strict Straightening Law

Some MSLs satisfy further natural axioms which strengthen (MSL-1) and (MSL-2). Let M be an MSL on $\mathcal{X}$ over A. The first additional axiom:

(MSL-3) For all $x, y \in \mathcal{X}$: $x < y \Rightarrow \mathcal{I}(x) \subset \mathcal{I}(y)$.

The property (MSL-3) implies that $\Pi \cup \mathcal{X}$ is a partially ordered set if we order its subsets Π and $\mathcal{X}$ as given and all other relations are given by

$$x < \xi \qquad \Longleftrightarrow \qquad \xi \notin \mathcal{I}(x).$$

(MSL-3) simply guarantees transitivity. If it is satisfied, one can consider the following strengthening of (MSL-2):

(MSL-4) $\xi x = \sum_{y < x, \xi} a_{\xi x y} y$ for all $x \in \mathcal{X}$, $\xi \in \mathcal{I}(x)$.

Definition. We say that M has a *strict straightening law* if it is an MSL satisfying (MSL-3) and (MSL-4).

An ideal $I \subset A$ generated by an ideal $\Psi \subset \Pi$ is a trivial example of a module with a strict straightening law, and the generic modules (3.3),(a) and (b) may be considered significant examples. On the other hand not every MSL has a strict straightening law. The following proposition which strengthens (2.7) excludes all the modules $M/I^n M$, $n \geq$ 2, as in (3.2), in particular the residue class rings $A/I^n A$, $n \geq 2$, $I = A\Psi$ straightening-closed.

(5.1) Proposition. *Let M be a module with a strict straightening law on $\mathcal{X}$ over A. Then*
$$\operatorname{Ann} M = A(\bigcap_{x \in \mathcal{X}} \mathcal{I}(x)).$$

PROOF: In fact, if $\xi \in \bigcap \mathcal{I}(x)$, then $\xi x = 0$ for all $x \in \mathcal{X}$, since there is no element $y \in \mathcal{X}$, $y < \xi$. —

Suppose that $\mathcal{X}$ is linearly ordered. Then the straightening laws (MSL-4) and (ASL-2) constitute a set of straightening relations on $\Pi \cup \mathcal{X}$, and the following question suggests itself: Is the symmetric algebra $S(M)$ an ASL over B? In general the answer is "no", as the following example demonstrates: $A = B[X_1, X_2, X_3]$, $X_1 < X_2 < X_3$,

$$M = A^3/(A(X_1, 0, 0) + A(X_2, 0, 0) + A(0, X_1, X_3)),$$

the residue classes of the canonical basis ordered by $\bar{e}_1 > \bar{e}_2 > \bar{e}_3$. On the other hand $S(I)$ is an ASL if I is generated by a linearly ordered poset ideal, cf. [BV.1, (9.13)] or [BST]; one uses that the Rees algebra $\mathcal{R}(I)$ of A with respect to I is an ASL, and concludes easily that the natural epimorphism $S(I) \to \mathcal{R}(I)$ is an isomorphism. We will give a new proof of this fact below.

The following proposition may not be considered ultima ratio, but it covers the case just discussed and also the generic modules.

(5.2) Proposition. *Let M be a graded module with strict straightening law on the linearly ordered set $\mathcal{X} = \{x_1, \ldots, x_n\}$, $x_1 < \cdots < x_n$. Put $\mathcal{X}_i = \{x_1, \ldots, x_i\}$, $M_i = A\mathcal{X}_i$, $\overline{M}_{i+1} = M/M_i$, $i = 0, \ldots, n$. Suppose that for all $j > i$ and all prime ideals $P \in \operatorname{Ass}(A/A\mathcal{I}(x_j))$ the localization $(\overline{M}_i)_P$ is a free $(A/A\mathcal{I}(x_i))_P$-module, $i = 1, \ldots, n$.*
(a) Then $S(M)$ is an ASL on $\Pi \cup \mathcal{X}$.
(b) If $\mathcal{I}(x_1) = \emptyset$, then $S(M)$ is a torsionfree A-module.

PROOF: Since $\Pi \cup \mathcal{X}$ generates $S(M)$ as a B-algebra (and $S(M)$ is a graded B-algebra in a natural way) and (ASL-2) is obviously satisfied, it remains to show that the standard monomials containing k factors from $\mathcal{X}$ are linearly independent for all $k \geq 0$. Since $S^0(M) = A$ this is obviously true for $k = 0$, and it remains true if $\operatorname{Ann} M = A\mathcal{I}(x_1)$ is factored out; since this does not affect the symmetric powers $S^k(M)$, $k > 0$, we may assume that $\operatorname{Ann} M = 0$. If $n = 1$, then M is now a free A-module and the contention holds for trivial reasons.

The hypotheses indicate that an inductive argument is in order. Independent of the special assumptions on M_i and $\mathcal{I}(x_i)$ there is an exact sequence

$$(*) \qquad S^k(M) \xrightarrow{g} S^{k+1}(M) \xrightarrow{f} S^{k+1}(M/Ax_1) \longrightarrow 0$$

in which f is the natural epimorphism and g is the multiplication by x_1. Let $P \in \operatorname{Ass} A$. By (2.8) x_1 generates a free direct summand of M_P. Therefore (5) splits over A_P, and $g \otimes A_P$ is injective. It is now enough to show that $S^k(M)$ is torsionfree; then g is injective itself and $(*)$ splits as a sequence of B-modules as desired: By induction the standard elements in $S^k(M)$ as well as in $S^{k+1}(M/Ax_1)$ are linearly independent.

The linear independence of the standard elements in $S^k(M)$ implies that $S^k(M)$ is an MSL over A on the set of monomials of length k in $\mathcal{X}$ with respect to a suitable partial order and the choice

$$\mathcal{I}(x_{i_1} \cdots x_{i_k}) = \mathcal{I}(x_{i_k}), \qquad i_1 \le \cdots \le i_k.$$

Let $P \in \operatorname{Spec} A$, $P \notin \operatorname{Ass} A$. Then $P \notin \operatorname{Ass}(A/A\mathcal{I}(x_1))$, since $\mathcal{I}(x_1) = \emptyset$ by assumption. If $P \notin \operatorname{Ass}(A/A\mathcal{I}(x_j))$ for all $j = 2, \ldots, n$, then $P \notin \operatorname{Ass} S^k(M)$ by virtue of (2.6); otherwise $S^k(M)_P$ is a free A_P-module by hypothesis. Altogether: $\operatorname{Ass} S^k(M) = \operatorname{Ass} A$, and $S^k(M)$ is torsionfree. —

(5.3) Corollary. *With the notations and hypotheses of* (5.2), *the symmetric algebra* $S(M_i)$ *is an ASL on* $\Pi \cup \mathcal{X}_i$ *for all* $i = 1, \ldots, n$. $S(M_i)$ *is a sub-ASL of* $S(M)$ *in a natural way.*

PROOF: There is a natural homomorphism $S(M_i) \to S(M)$ induced by the inclusion $M_i \to M$. Since $S(M_i)$ satisfies (ASL-2), it is generated as a B-module by the standard monomials in $\Pi \cup \mathcal{X}_i$. Since these standard monomials are linearly independent in $S(M)$, they are linearly independent in $S(M_i)$, too, and $S(M_i)$ is a subalgebra of $S(M)$. —

The following corollary has already been mentioned:

(5.4) Corollary. *Let A be an ASL on Π, and $\Psi \subset \Pi$ a linearly ordered ideal. Then $S(A\Psi)$ is an ASL on the disjoint union of Π and Ψ.*

PROOF: For each $\psi \in \Psi$ the poset $\Pi \setminus \mathcal{I}(\psi)$ has ψ as its single minimal element. Let $\Psi = \{\psi_1, \ldots, \psi_n\}$, $\psi_1 < \cdots < \psi_n$. If $P \in \operatorname{Ass}(A/A\mathcal{I}(\psi_j))$, then $\psi_j \notin P$ since ψ_j is not a zero-divisor of the ASL $A/A\mathcal{I}(\psi_j)$. Consequently $(A\Psi/(\sum_{k=1}^i A\psi_k))_P$ is isomorphic to $(A/\mathcal{I}(\psi_i))_P$ for all $i < j$. —

We want to apply (5.2) to the generic modules discussed in (3.3), (a), and recall the notations introduced there: $A = B[X]/I_{r+1}(X)$ is an ASL on $\Delta_r(X)$, the set of all i-minors, $i \le r$, of X. M is the cokernel of the map $A^m \to A^n$ defined by the matrix x, $\bar{e}_1, \ldots, \bar{e}_n$ are the residue classes of the canonical basis $e_1, \ldots, e_n$ of A^n. (Thus M_k is the submodule of M generated by $\bar{e}_{n-k+1}, \ldots, \bar{e}_n$.)

(5.5) Corollary. (a) *With the notations just recalled, the symmetric algebra of a generic module M is an ASL. If $r + 1 \le n$, $S(M)$ is torsionfree over A.*
(b) *Let B be a Cohen-Macaulay ring. $S(M)$ is Cohen-Macaulay if and only if $r + 1 \le n$ or $r = m = n$.*

PROOF: (a) Factoring out the ideal generated by $\mathcal{I}(\bar{e}_n)$ we may suppose that $r < n$. Note that with the notations introduced in (3.3),(a) one has $\bar{e}_n < \cdots < \bar{e}_1$. Because of statement (ii) in (3.3),(a) the validity of the hypothesis of (5.2) for $i \ge n - r + 1$ follows from the proof of (5.4).

Let $i \le n - r$, $j > i$, $k = n - j + 1$, $\delta = [1, \ldots, r | 1, \ldots, r]$ for $k \ge r + 1$ and $\delta = [1, \ldots, r | 1, \ldots, \hat{k}, \ldots, r + 1]$ for $k \le r$. Then δ is the minimal element of the poset underlying $A/\mathcal{I}(x_j) = A/\mathcal{I}(\bar{e}_k)$, thus not contained in an associated prime ideal of the latter. On the other hand $(\overline{M}_i)_P$ is free for every prime P not containing δ.

(b) in order to form the poset $\Pi \cup \{\bar{e}_1, \ldots, \bar{e}_n\}$ one attaches $\{\bar{e}_1, \ldots, \bar{e}_n\}$ to Π as indicated by the following diagrams for the cases $r+1 \le n$ and $r = m = n$ resp. In the first case we let $\delta_i = [1, \ldots, r|1, \ldots, \hat{i}, \ldots, r+1]$, in the second $\delta_i = [1, \ldots, r-1|1, \ldots, \hat{i}, \ldots, r]$.

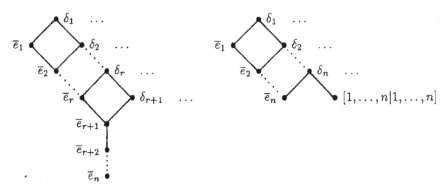

It is an easy exercise to show that $\Pi \cup \{\bar{e}_1, \ldots, \bar{e}_n\}$ and $\Pi \cup \{\bar{e}_{n-k+1}, \ldots, \bar{e}_n\}$ are wonderful, implying the Cohen-Macaulay property for ASL's defined on the poset ([BV.1, Section 5] or [DEP.2]).

In the case in which $m > n = r$, the ideal $I_n(X) S(M)$ annihilates $\bigoplus_{i>0} S^i(M)$, and $\dim S(M)/I_n(X) < \dim S(M)$ by (1.3), excluding the Cohen-Macaulay property. —

Admittedly the preceding corollary is not a new result. In fact, let Y be an $n \times 1$ matrix of new indeterminates. Then

$$S(M) \cong B[X, Y]/(I_{r+1}(X) + I_1(XY))$$

can be regarded as the coordinate ring of a variety of complexes, which has been shown to be a Hodge algebra in [DS]. The results of [DS] include part (b) of (5.5) as well as the fact that $S(M)$ is a (normal) domain if $r + 1 \le n$ and B is a (normal) domain. The divisor class group of $S(M)$ in case $r + 1 \le b$, B normal, has been computed in [Br.2]: $\mathrm{Cl}(S(M)) = \mathrm{Cl}(B)$ if $m = r < n - 1$, $\mathrm{Cl}(S(M)) = \mathrm{Cl}(B) \oplus \mathbf{Z}$ else. The algebras $S(M)$, in particular for the cases $r + 1 > \min(m, n)$, i.e. $A = B[X]$, and $r + 1 = \min(m, n)$, have received much attention in the literature, cf. [Av.2], [BE], [BKM], and the references given there. Note that (5.5) also applies to the subalgebras $S(M_k)$. In the case $A = B[X]$, $m \le n$, these rings have been analyzed in [BS].

The analogue (5.6) of (5.5) seems to be new however. We recall the notations of (3.3),(b): X is an alternating $n \times n$-matrix of indeterminates, $A = B[X]/\mathrm{Pf}_{r+2}(X)$, $F = A^n$, $x: F \to F^*$ given by the residue class of X, and $M = \mathrm{Coker}\, x$.

(5.6) Corollary. (a) *With the notations just recalled, the symmetric algebra of an "alternating" generic module M is an ASL. If $r < n$, $S(M)$ is a torsionfree A-module.*
(b) *Let B be a Cohen-Macaulay ring. Then $S(M)$ is Cohen-Macaulay if and only if $r < n$.*
(c) *Let B be a (normal) domain. Then $S(M)$ is a (normal) domain if and only if $r < n$.*
(d) *Let B be normal and $r < n$. Then $\mathrm{Cl}(S(M)) \cong \mathrm{Cl}(B) \oplus \mathbf{Z}$ if $r = n - 1$, and $\mathrm{Cl}(S(M)) \cong \mathrm{Cl}(B)$ if $r < n - 1$. In particular $S(M)$ is factorial if $r < n - 1$ and B is factorial.*

PROOF: (a) and (b) are proved in the same way as (5.5).

Standard arguments involving flatness reduce (c) to the case in which B is a field (cf. [BV.1, Section 3] for example). Thus we may certainly suppose that B is a normal domain.

In the case in which $r = n - 1$ the module M is just $I = \mathrm{Pf}_{n-1}(X)$ as remarked above, an ideal generated by a linearly ordered poset ideal. Then (i) $\mathrm{Gr}_I A$ is an ASL, in particular reduced, and (ii) $S(M)$ is the Rees algebra of A with respect to I (cf. [BST] for example). Thus we can apply the main result of [HV] to conclude (c) and (d).

Let $r \leq n - 2$ now. In the spirit of this paper a "linear" argument seems to be most appropriate: By [Fo, Theorem 10.11] and [Av.1] it is sufficient that all the symmetric powers of M are reflexive. Since M_P, hence $S^k(M_P)$ is free for prime ideals $P \not\supseteq \mathrm{Pf}_r(x)$ it is enough to show that $\mathrm{Pf}_r(x)$ contains an $S^k(M)$-sequence of length 2 for every k. Each $S^k(M)$ is an MSL whose data $\mathcal{I}(\dots)$ coincide with those of M itself. Therefore (2.6) can be applied and we can replace the $S^k(M)$ by the residue class rings A/I_i, $I_i = A\{\pi \in \Phi_r(x) \colon \pi \not\geq [1,\dots,\widehat{i},\dots,r+1]\}$, $i = 1,\dots,r$. One has $\mathrm{Pf}_r(X) \supset I_i$.

The poset Π underlying A/I_i is wonderful (cf. [DEP.2, Lemma 8.2] or [BV.1, (5.13)]). Therefore the elements

$$[1,\dots,\widehat{i},\dots,r+1] = \sum_{\substack{\pi \in \Pi \\ rk\,\pi = 1}} \pi \qquad \text{and} \qquad \sum_{\substack{\pi \in \Pi \\ rk\,\pi = 2}} \pi$$

form an A/I_i-sequence by [DEP.2, Theorem 8.1]. Both these elements are contained in $\mathrm{Pf}_r(x)$.

References

[AD] ABEASIS, S., DEL FRA, A., *Young diagrams and ideals of Pfaffians*, Adv. Math. **35** (1980), 158–178.

[Av.1] AVRAMOV, L.L., *A class of factorial domains*, Serdica **5** (1979), 378–379.

[Av.2] AVRAMOV, L.L., *Complete intersections and symmetric algebras*, J. Algebra **73** (1981), 248–263.

[Br.1] BRUNS, W., *Generic maps and modules*, Compos. Math. **47** (1982), 171–193.

[Br.2] BRUNS, W., *Divisors on varieties of complexes*, Math. Ann. **264** (1983), 53–71.

[Br.3] BRUNS, W., *Addition to the theory of algebras with straightening law*, in: M. Hochster, C. Huneke, J.D. Sally (Ed.), Commutative Algebra, Springer 1989.

[BKM] BRUNS, W., KUSTIN, A., MILLER, M., *The resolution of the generic residual intersection of a complete intersection*, J. Algebra (to appear).

[BS] BRUNS, W., SIMIS, A., *Symmetric algebras of modules arising from a fixed submatrix of a generic matrix.*, J. Pure Appl. Algebra **49** (1987), 227–245.

[BST] BRUNS, W., SIMIS, A., NGÔ VIỆT TRUNG, *Blow-up of straightening closed ideals in ordinal Hodge algebras*, Trans. Amer. Math. Soc. (to appear).

[BV.1] BRUNS, W., VETTER, U., "Determinantal rings," Springer Lect. Notes Math. **1327**, 1988.

[BV.2] BRUNS, W., VETTER, U., *Modules defined by generic symmetric and alternating maps*, Proceedings of the Minisemester on Algebraic Geometry and Commutative Algebra, Warsaw 1988 (to appear).

[BE] BUCHSBAUM, D., EISENBUD, D., *Algebra structures for finite free resolutions, and some structure theorems for ideals of codimension 3*, Amer. J. Math. **99** (1977), 447-485.

[DEP.1] DE CONCINI, C., EISENBUD, D., PROCESI, C., *Young diagrams and determinantal varieties*, Invent. Math. **56** (1980), 129-165.

[DEP.2] DE CONCINI, C., EISENBUD, D., PROCESI, C., "Hodge algebras," Astérisque **91**, 1982.

[DS] DE CONCINI, C., STRICKLAND, E., *On the variety of complexes*, Adv. Math. **41** (1981), 57-77.

[Ei] EISENBUD, D., *Introduction to algebras with straightening laws*, in "Ring Theory and Algebra III," M. Dekker, New York and Basel, 1980, pp. 243-267.

[EH] EISENBUD, D., HUNEKE, C., *Cohen-Macaulay Rees algebras and their specializations*, J. Algebra **81** (1983), 202-224.

[Fo] FOSSUM, R.M., "The divisor class group of a Krull domain," Springer, Berlin - Heidelberg - New York, 1973.

[Go] GOTO, S., *On the Gorensteinness of determinantal loci*, J. Math. Kyoto Univ. **19** (1979), 371-374.

[HK] HERZOG, J., KUNZ, E., "Der kanonische Modul eines Cohen-Macaulay-Rings," Springer Lect. Notes Math. **238**, 1971.

[HV] HERZOG, J., VASCONCELOS, W.V., *On the divisor class group of Rees algebras*, J. Algebra **93** (1985), 182-188.

[Hu] HUNEKE, C., *Powers of ideals generated by weak d-sequences*, J. Algebra **68** (1981), 471-509.

[KL] KLEPPE, H., LAKSOV, D., *The algebraic structure and deformation of Pfaffian schemes*, J. Algebra **64** (1980), 167-189.

[Ku] KUTZ, R.E., *Cohen-Macaulay rings and ideal theory of invariants of algebraic groups*, Trans. Amer. Math. Soc. **194** (1974), 115-129.

[Ma] MATSUMURA, H., "Commutative Algebra," Second Ed., Benjamin/Cummings, Reading, 1980.

[Ve.1] VETTER, U., *The depth of the module of differentials of a generic determinantal singularity*, Commun. Algebra **11** (1983), 1701-1724.

[Ve.2] VETTER, U., *Generische determinantielle Singularitäten: Homologische Eigenschaften des Derivationenmoduls*, Manuscripta Math. **45** (1984), 161-191.

Universität Osnabrück, Abt. Vechta, Driverstr. 22, D-2848 Vechta

On short graded algebras

Maria Pia Cavaliere, Maria Evelina Rossi, and Giuseppe Valla

Introduction

Let (A, m, k) be a local Cohen-Macaulay ring of dimension d. We denote by e the multiplicity of A, by N its embedding dimension and by $h := N - d$ the codimension of A. The Hilbert function of A is the numerical function defined by $H_A(n) := \dim_k(m^n/m^{n+1})$ and the Poincare series is the series $P_A(z) := \sum_{n \geq 0} H_A(n)z^n$. By the theorem of Hilbert-Serre there exists a polynomial $f(z) \in \mathbf{Z}[z]$ such that $f(1) = e$ and $P_A(z) = f(z)/(1-z)^d$. From this it follows that there exists a polynomial $h_A(x) \in \mathbf{Q}[x]$ such that $H_A(n) = h_A(n)$ for all $n \gg 0$. This polynomial is called the Hilbert polynomial of A. If we denote by $s = s(A) := \deg(f(z))$ and by $i = i(A) := \max\{n \in \mathbf{Z} | H_A(n) \neq h_A(n)\} + 1$, then it is well known that $i = s - d + 1$ (see [EV]). Also we denote by $t = t(A)$ the initial degree of A, which is by definition $t = t(A) := \min\{j | H_A(j) \neq \binom{N+j-1}{j}\}$. It is clear from the definition that $t \geq 2$. In [RV] we proved that $e \geq \binom{h+t-1}{h}$. Also in the same paper we proved that if $e = \binom{h+t-1}{h}$ then $\mathrm{gr}_m(A) := \oplus(m^n/m^{n+1})$ is a Cohen-Macaulay graded ring and

$$P_A(z) = \sum_{i=0}^{t-1} \binom{h+i-1}{i} z^i/(1-z)^d.$$

If $e = \binom{h+t-1}{h} + 1$ then $\mathrm{gr}_m(A)$ needs not to be Cohen-Macaulay (see [S]) but if the Cohen-Macaulay type $\tau(A)$ verifies $\tau(A) < \binom{h+t-2}{t-1}$ then again $\mathrm{gr}_m(A)$ is Cohen-Macaulay and

$$P_A(z) = \left(\sum_{i=0}^{t-1} \binom{h+i-1}{i} z^i + z^t \right) /(1-z)^d.$$

(see [RV]). On the other hand if we consider a set X of e distinct points in the projective space $\mathbf{P}^h$ and we let $A = k[X_0, \ldots, X_h]/I$ be the coordinate ring of X, then A is a graded Cohen-Macaulay ring of dimension 1. Hence the Hilbert function of A is strictly increasing up to the degree of X, which is e. Many authors (see [GO1],[G],[GO2],[GM],[GGR], [B],[Br1],[Br2],[BK],[L1],[L2],[R],[TV]) have studied the notion of points in "generic" position. This means by definition that

$$H_A(n) = \min\left(e, \binom{h+n}{n} \right).$$

It is easy to prove that almost every set of e points in $\mathbf{P}^h$ are in generic position, in the

The first two authors were partially supported by M.P.I. (Italy). The third author thanks the Max-Planck-Institut für Mathematik in Bonn for hospitality and financial support during the preparation of this paper.

sense that the points in generic position in $\mathbf{P}^h$ form a dense open set U of $\mathbf{P}^h \times \mathbf{P}^h \times \cdots \times \mathbf{P}^h$ (e times). Now it is clear that if X is a set of points in generic position in $\mathbf{P}^h$ then

$$P_A(z) = \left(\sum_{i=0}^{t-1} \binom{h+i-1}{i} z^i + cz^t \right) / (1-z)$$

where t is defined to be the integer such that $\binom{h+t-1}{h} \le e < \binom{h+t}{h}$.

Thus we are led to consider graded algebras $A = k[X_0, \ldots, X_r]/I$ over an infinite field k which are Cohen-Macaulay and whose Poincare series is given by

$$P_A(z) = \left(\sum_{i=0}^{t-1} \binom{h+i-1}{i} z^i + cz^t \right) / (1-z)^d$$

where d is the Krull dimension of A, t is an integer ≥ 2, and c is an integer $0 \le c < \binom{h+t-1}{t}$.

We call such an algebra a **Short Graded Algebra**.

It is easy to see that short graded algebras are the Cohen-Macaulay graded algebras A such that H_A^{1-d} is maximal according to the definition given by Orecchia in [O] . Also extremal Cohen-Macaulay graded algebras in the sense of Schenzel (see [Sc]) are short graded algebras with $c = 0$.

Generalities on short graded algebras

Let $A = k[X_0, \ldots, X_r]/I$ be a short graded algebra with Poincare series

$$P_A(z) = \left(\sum_{i=0}^{t-1} \binom{h+i-1}{i} z^i + cz^t \right) / (1-z)^d.$$

The multiplicity of A is denoted by $e = e(A)$. We have $e = \binom{h+t-1}{h} + c$. Also we have $i = i(A) = t - d + 1$. Since k is an infinite field, we can find d linear forms $L_1, \ldots, L_d$ in $R = k[X_0, \ldots, X_r]$ such that if $J = (L_1, \ldots, L_d)$, the graded algebra $B = A/JA$ is of dimension 0, codimension h and has $e(A) = e(B)$. If we denote by - reduction modulo J, we get $B = \bar{R}/\bar{I}$ and we call B an artinian reduction of A. It is clear that B is a short graded algebra with

$$P_B(z) = \sum_{i=0}^{t-1} \binom{h+i-1}{i} z^i + cz^t.$$

It follows that $s(B) = s(A) = t(B) = t(A) = t$. Now let

$$\mathbf{F} : 0 \to \bar{F}_h \to \cdots \to \bar{F}_1 \to \bar{R} \to B \to 0$$

be a minimal graded free resolution of B with $\bar{F}_i = \oplus_{j=1}^{\beta_i} \bar{R}(-d_{ij})$. The positive integers β_i are called the Betti numbers of B; the integers d_{ij} are called the shifting in the resolution of B and , along with the β_i, are unique. Since $t(B) = t$ we have $t \le d_{1j}$ for every j. Further it is well known that we have a graded isomorphism $\text{Tor}_h^{\bar{R}}(B, k) \simeq (0 : B_1)(-h)$, hence we get $d_{hj} \le s + h$ for every j. The following lemma is possibly well known, but we insert here a proof for the sake of completeness.

Let

$$\mathbf{F} : 0 \to \bar{F}_h \to \bar{F}_{h-1} \to \cdots \to \bar{F}_0 \to M \to 0$$

be a minimal graded free resolution of the graded R-module M, with $F_i = \oplus_{j=1}^{\beta_i} R(-d_{ij})$.

Lemma 1.1. *If $i > 0$, for every j there exists q such that $d_{i-1,q} < d_{ij}$. If $i < h$, for every j there exists p such that $d_{ij} < d_{i+1,p}$.*

PROOF: It is clear that d_{ij} is the degree of the element of F_{i-1} which is the j-th column of the matrix Δ_i representing the map of free modules $F_i \to F_{i-1}$. Hence we get for every $q = 1, \dots, \beta_{i-1}$

$$\delta_q + d_{i-1,q} = d_{ij}$$

where $\delta_1, \dots, \delta_{\beta_{i-1}}$ are the degree of the elements of this column vector. Now if for some j we have $d_{ij} = d_{i-1,q}$ for every q, then Δ_i would have a column of zeros, a contradiction to the minimality of the resolution. The other result follows in the same way, by using the fact that the transpose of Δ_i cannot have a column of zeros since it is a matrix in the minimal graded free resolution of $\operatorname{Ext}_R^h(M, R)$.

Using this lemma we get that in the resolution $\bar{\mathbf{F}}$ of B we have

$$\bar{F}_i = \bar{R}(-t - i)^{b_i} \oplus \bar{R}(-t - i + 1)^{a_i}$$

for every $i \geq 1$. Now it is well known that the graded free resolution of A as an R-module has the same Betti numbers and shifting as the resolution of B as an $\bar{R}$-module. Hence a graded free resolution of A can be written as

$$0 \to R(-t - h)^{b_h} \oplus R(-t - h + 1)^{a_h} \to \dots$$
$$\to R(-t - i)^{b_i} \oplus R(-t - i + 1)^{a_i} \to \dots \to R \to A \to 0$$

for some integers $a_i, b_i \geq 0$. By the particular Hilbert function of A we get $a_1 = \binom{h+t-1}{t} - c$ and $b_h = c$.

A detailed proof of these observations can be found in [L2].

We close this section by remarking that for a short graded algebra the Betti numbers β_i determine all the resolution. This can be easily seen by using the fact that in each degree $n > t$ we have

$$\dim(\bar{R}_n) + \sum_{i=1}^h (-1)^i \left[a_i \dim(\bar{R}(-t - i + 1)_n) + b_i \dim(\bar{R}(-t - i)_n) \right] = 0.$$

Pure and linear resolution

Recall that given a graded free resolution

$$\mathbf{F} : 0 \to F_h \to \dots \to F_1 \to R \to A \to 0$$

of the graded algebra A with $F_i = \oplus_{j=1}^{\beta_i} R(-d_{ij})$ we say that the resolution is pure of type $(d_1, \dots, d_h)$ if for every $i = 1, \dots, h$ we have $d_{ij} = d_i$ for every j. If the resolution is pure of type $(t, t + m, t + 2m, \dots, t + (h - 1)m)$, we shall say that it is pure of type (t, m). A pure resolution of type $(t, 1)$ is just called a t-linear resolution (see [W],[HK])

In this section we investigate what short graded algebras have pure or linear resolution. The first proposition deals with the case of a linear resolution.

Proposition 2.1. *Let A be a Cohen-Macaulay graded algebra. The following conditions are equivalent*
   a) A is short and has a t-linear resolution.
   b) A is short with $c = 0$.
   c) $e = \binom{h+t-1}{h}$ and $t = \mathrm{indeg}(A)$
   d) I is generated by $\binom{h+t-1}{t}$ forms of degree t

PROOF: The conditions b), c) and d) are equivalent by theorem 3.3 in [RV]. If A is short and $c = 0$ then $b_h = 0$. By lemma 1.1 this implies $b_i = 0$ for every $i = 1, \ldots, h$ and the resolution is linear. If the resolution is linear then $b_h = 0$, hence $c = 0$.

The case of a pure resolution of type (t, m) is considered in the next proposition which extends Theorem 2 in [Br1].

Proposition 2.2. *Let A be a short graded algebra. A has a pure resolution of type (t, m) if and only if one of the following occurs*
   a) $e = \binom{h+t-1}{h}$
or
   b) $h = 2$, $e = \binom{t+1}{2} + \frac{t}{2}$ where t is even and I is generated by forms of degree t.

PROOF: If the resolution is linear a) holds by the above proposition. If the resolution is pure of type (t, m) with $m \geq 2$, we get $d_h = t + (h - 1)m \leq t + h$, hence $(h - 1)m \leq h$. This implies $m = 1$ or $m = h = 2$. In the first case a) holds by the above proposition, while in the latter case we get a resolution

$$0 \to R(-t-2)^{a-1} \to R(-t)^a \to R \to A \to 0$$

From this it follows easily that t is even, $e = \binom{t+1}{2} + \frac{t}{2}$ and I is generated by forms of degree t.

Conversely if a) holds the conclusion follows by the above proposition, while if b) holds we get a resolution

$$0 \to R(-t-2)^{b_2} \oplus R(-t-1)^{a_2} \to R(-t)^{a_1} \to R \to A \to 0$$

It follows that $b_2 + a_2 = a_1 - 1$ where $a_1 = t + 1 - c$ and $b_2 = c = \frac{t}{2}$. Hence $a_2 = t + 1 - \frac{t}{2} - 1 - \frac{t}{2} = 0$

The next result says that a short graded algebra has a pure resolution if and only if it has some special Betti numbers. It extends Theorem 3 in [Br1] (see also [L1]).

Proposition 2.3. *Let A be a short graded algebra with $\binom{h+t-1}{h} < e < \binom{h+t}{h}$. A has a pure resolution if and only if there exists an integer p such that $1 \leq p \leq h - 1$ and*

$$\beta_i = \begin{cases} \binom{t+i-2}{i-1}\binom{h+t}{h-i+1}\frac{p-i+1}{t+p}, & \text{for } i = 1, \ldots, p, \\ \binom{t+i-1}{i}\binom{h+t}{h-i}\frac{i-p}{t+p}, & \text{for } i = p+1, \ldots, h. \end{cases}$$

PROOF: If A is short and has a pure resolution of type $(d_1, \ldots, d_h)$, then $d_1 = t$ and $d_h = t + h$, otherwise if $d_h = t + h - 1$ then the resolution would be linear and by Proposition 2.1 $e = \binom{h+t-1}{h}$. Hence there exists an integer p, $1 \leq p \leq h - 1$ such that

$$d_i = \begin{cases} t + i - 1, & \text{for } i = 1, \ldots, p, \\ t + i, & \text{for } i = p+1, \ldots, h. \end{cases}$$

Now, by a result of Herzog and Kuhl (see [HK]), if the graded algebra A has a pure resolution of type $(d_1, \ldots, d_h)$ then $\beta_i = \left| \prod_{j \neq i} \frac{d_j}{d_j - d_i} \right|$. In our case the conclusion follows by an easy computation. Conversely, we have seen at the end of section 1 that for a short graded algebra the Betti numbers determine all the resolution. Now it is easy to prove that the particular Betti numbers of the proposition determine a pure resolution.

For example let us consider the case $h = 3$, $t = 3$, $p = 2$. We get $\beta_1 = 8$, $\beta_2 = 9$, $\beta_3 = 2$, hence we have a resolution

$$\leq 0 \to R(-6)^{b_3} \oplus R(-5)^{a_3} \to R(-5)^{b_2} \oplus R(-4)^{a_2} \to R(-4)^{b_1} \oplus R(-3)^{a_1} \to R \to A \to 0$$

with $a_1 = 10 - c$, hence $b_1 = c - 2$. Now $b_3 = c \leq \beta_3 = 2$, hence $c = 2$, $a_1 = 8$, $b_1 = 0$, $b_3 = 2$, $a_3 = 0$. Further we have

$$\dim(\bar{R}_4) + a_2 = b_1 + a_1 \dim(\bar{R}_1).$$

Since $b_1 = 0$, $\dim(\bar{R}_4) = \binom{3+4-1}{4} = 15$, $\dim(\bar{R}_1) = 3$ we get $a_2 = 9$, hence $b_2 = 0$ and the resolution is pure of type $(3, 4, 6)$.

We finally remark that if A is a short graded algebra with a pure resolution, then for the same p as in the above proposition, we get $e = \frac{t\binom{h+t}{h}}{t+p}$ (see [HM]).

A particular case of pure resolution is considered in the last result of this section.

Theorem 2.4. *Let $A = R/I$ be a graded algebra which is Cohen-Macaulay. Then the following conditions are equivalent:*

a) A is Gorenstein and short.

b) A has a pure resolution and $e = h + 2$.

c) The resolution of A is

$$0 \to R(-h-2)^{\beta_h} \to R(-h)^{\beta_{h-1}} \to \cdots \to R(-2)^{\beta_1} \to R \to A \to 0$$

PROOF: If A is Gorenstein the Hilbert function of its artinian reduction is symmetric, hence we get $c = 1$, $e = h + 2$ and $t = 2$. This proves that A is an extremal Gorenstein algebra according to the definition given by Schenzel in [Sc]. But extremal Gorenstein algebras have a pure resolution of type $(2, 3, \ldots, h, h + 2)$ as proved in the same paper [Sc]. Hence a) implies b) and c). Let now prove that b) implies c). It is clear that $P_A(z) = (1 + hz + z^2)/(1 - z)^d$, hence $c = 1$ and $b_h = c = 1$. Since the resolution is pure we get $\beta_h = 1$ and A is Gorenstein. Finally we prove that c) implies a). By the formula of Herzog and Kuhl we get

$$\beta_h = \left| \prod_{j < h} \frac{d_j}{d_j - h - 2} \right| = \left| \frac{2}{-h} \frac{3}{-h+1} \cdots \frac{h}{-2} \right| = \frac{h!}{h!} = 1$$

hence A is Gorenstein. Further I is generated by forms of degree 2 and we get

$$\beta_1 = a_1 = \left| \prod_{j > 1} \frac{d_j}{d_j - 2} \right| = \left| \frac{3}{1} \frac{4}{2} \cdots \frac{h}{h-2} \frac{h+2}{h} \right| = \frac{h!(h+2)}{2(h-2)!h} = \binom{h+1}{2} - 1.$$

The conclusion follows by using theorem 3.10 in [RV].

Right almost linear resolution

Let A be a graded algebra with graded free resolution

$$\mathbf{F} : 0 \to F_h \to F_{h-1} \to \cdots \to F_1 \to R \to A \to 0$$

where $F_i = \oplus_{j=1}^{\beta_i} R(-d_{ij})$. Following [L1] we say that $\mathbf{F}$ is right almost linear if it is linear except possibly at F_1. In [L1] Lorenzini proved that the coordinate ring of a set of points in $\mathbf{P}^h$ has a right almost linear resolution in some particular cases. All these results are consequence of the following theorem which proves that a suitable condition on the defining ideal of a short graded algebra forces the resolution to be right almost linear with special Betti numbers.

We recall that for a short graded algebra $A = R/I$, N denotes the embedding dimension of A. Hence we may assume $A = R/I$ where R is a polynomial ring of dimension N. As before we let $B = \bar{R}/\bar{I}$ be an artinian reduction of A. (see section 1).

Theorem 3.1. *Let A be a short graded algebra such that $e = \binom{h+t}{h} - p$ for some positive integer p. If $\dim_k(I_t R_1) = Np$ then the resolution of A is right almost linear of type*

$$0 \to R(-t-h)^{b_h} \to \cdots \to R(-t-2)^{b_2} \to R(-t-1)^{b_1} \oplus R(-t)^{a_1} \to R \to A \to 0$$

where $a_1 = p$, $b_1 = \binom{h+t}{h-1} - hp$, $b_i = \binom{h}{i}e - \binom{i+t-1}{i}\binom{h+t}{h-i}$ for every $i = 2, \ldots, h$.

PROOF: Since $e = \binom{h+t}{h} - p = \binom{h+t-1}{h} + \binom{h+t-1}{t} - p$ we get $c = \binom{h+t-1}{t} - p$, hence $a_1 = p$. This means $\dim_k(I_t) = p$, and since $\dim_k(I_t R_1) = Np$ we get $a_2 = 0$. By lemma 1.1 this implies $a_i = 0$ for every $i \geq 2$. Since in each degree $n > t$ we have

$$\dim(\bar{R}_n) + \sum_{i=1}^{h}(-1)^i \left[a_i \dim(\bar{R}(-t-i+1)_n) + b_i \dim(\bar{R}(-t-i)_n) \right] = 0.$$

we get $\dim(\bar{R}_{t+1}) - a_1 \dim(\bar{R}_1) - b_1 = 0$, hence $b_1 = \binom{h+t}{t+1} - ph$. In the same way we get $\dim(\bar{R}_{t+2}) - a_1 \dim(\bar{R}_2) - b_1 \dim(\bar{R}_1) + b_2 = 0$, from which, by easy computation, one gets $b_2 = \binom{h}{2}e - \binom{t+1}{2}\binom{h+t}{h-2}$. By induction we get the right value of the remaining b_i's.

We remark that we can apply the above results to the following cases:
a) $e = \binom{h+t}{h} - 1$ points in generic position in $\mathbf{P}^h$
b) $e = \binom{h+t}{h} - 2$ points in uniform position in $\mathbf{P}^h$.

In fact in case a) I_t is a vector space of dimension 1, hence it is clear that the condition of the theorem is fullfilled. As for the case b) we recall that a set of e points in $\mathbf{P}^h$ is said to be in uniform position if every subset is in generic position. Now case b) follows from the following lemma a stronger version of which has been proved by Geramita and Maroscia in [GM] by completely different methods. We insert here a proof since the original one is rather complicate.

As usual we denote by $A = k[X_0, \ldots, X_n]/I$ the coordinate ring of a set of points in $\mathbf{P}^h$ and by t the initial degree of A.

Lemma 3.2. *If $P_1, \ldots, P_e$ are points in uniform position in $\mathbf{P}^h$, the forms of degree t in I cannot have a common factor (if $\dim(I_t) = 1$ and $I_t = kF$ this means that F is irreducible).*

PROOF: Let F be a common factor of all the forms in I_t with $\deg(F) = d$, $1 \leq d \leq t - 1$. Let $\wp_1, \ldots, \wp_e$ be the prime ideals of the poits $P_1, \ldots, P_e$ respectively. Since $d < t = \text{indeg}(A)$ we must have $F \in \wp_1 \cap \cdots \cap \wp_n$, $F \notin \wp_{n+1} \cup \cdots \cup \wp_e$ for some n, $1 \leq n < e$. Let $K = \wp_1 \cap \cdots \cap \wp_n$, $J = \wp_{n+1} \cap \cdots \cap \wp_e$. It is clear that $I_t = FJ_{t-d}$, hence $\dim(I_t) = \dim(J_{t-d})$ and we get $H_{R/J}(t - d) = \binom{h+t-d}{h} - \dim(I_t)$. Since $P_{n+1}, \ldots, P_e$ are in generic position we have $H_{R/J}(t - d) = \min \left\{ e - n, \binom{h+t-d}{h} \right\}$, hence we get $e - n = \binom{h+t-d}{h} - \dim(I_t) = \binom{h+t-d}{h} - \binom{h+t}{h} + H_{R/I}(t) \leq \binom{h+t-d}{h} - \binom{h+t}{h} + e$. This implies $n \geq \binom{h+t}{h} - \binom{h+t-d}{h} \geq \binom{h+d}{h}$ where the last inequality follows by an easy combinatorial argument. Thus we get $H_{R/K}(d) = \min \left\{ n, \binom{h+d}{h} \right\} = \binom{h+d}{h}$, a contradiction to the fact that $F \in K$.

The Cohen-Macaulay type

In this section we study the Cohen-Macaulay type of some special classes of short graded algebras. The first theorem extends and simplifies analogous results given by Brown and Roberts (see [Br2] and [R]).

Theorem 4.1. *Let A be a short graded algebra with $e = \binom{h+t}{h} - p$ for some positive integer p. Let J be the ideal generated by the forms of degree t in I. If $h(J) > p - h + 1$ then $\beta_h = \binom{h+t-1}{t} - p$*

PROOF: Since k is an infinite field, it is clear that given a maximal regular sequence of forms of degree t in I we may complete this to a maximal regular sequence in R with linear forms $L_1, \ldots, L_d$ such that $A/(L_1, \ldots, L_d)A = \bar{R}/\bar{I}$ is an artinian reduction of A. Hence $h(J)$ coincides with the height of the corresponding ideal generated by the forms of degree t in $\bar{I}$. Thus we may assume $A = k[X_1, \ldots, X_h]/I$ with $\dim(A) = 0$. We have $b_h = c = \binom{h+t-1}{t} - p$, hence we need only to prove that $a_h = 0$, or which is the same, that if F is a form of degree $t - 1$ such that $FR_1 \subseteq I$, then $F = 0$. We have $\dim(I_t) = p$, hence if $p < h$ the conclusion is clear. Let $p \geq h$ and F be a form of degree $t-1$ such that $FR_1 \subseteq I$. Then $FX_1, \ldots, FX_h$ are linearly independent vectors in I_t, hence we can find vectors $G_1, \ldots, G_{p-h} \in I_t$ such that $(FX_1, \ldots, FX_h, G_1, \ldots, G_{p-h})$ is a k-vector base of I_t. This means that $J \subseteq (F, G_1, \ldots, G_{p-h})$, hence $h(J) \leq p - h + 1$, a contradiction.

The case of e points in generic position in $\mathbf{P}^h$ with $e = \binom{h+t}{h} - p$ and $p \leq h - 1$ is the main result in [R].

On the other hand if we have $e = \binom{h+t}{h} - h$ points in uniform position, by lemma 3.2 we get $h(J) \geq 2$ and we may apply the above theorem. This is the main result in [Br2].

Let now $A = R/I$ be a Cohen-Macaulay graded algebra with codimension h, multiplicity e and initial degree t. It is clear that $e \geq \binom{h+t-1}{h}$ and we have seen in proposition 2.1 that if $e = \binom{h+t-1}{h}$ then A is short and the resolution is t-linear. In the following proposition we study the case $e = \binom{h+t-1}{h} + 1$.

Proposition 4.2. *Let A be a Cohen-Macaulay graded algebra with $e = \binom{h+t-1}{h} + 1$. Then we have:*

a) A is short with $c = 1$.

b) $\beta_h \leq \binom{h+t-2}{t-1}$.

c) The following condition are equivalent:

  c1) $\beta_h < \binom{h+t-2}{t-1}$

  c2) $b_1 = 0$

  c3) $\beta_1 = \binom{h+t-1}{t} - 1$

d) The following conditions are equivalent:

  d1) $\beta_h = \binom{h+t-2}{t-1}$

  d2) $b_1 = 1$

  d3) $\beta_1 = \binom{h+t-1}{t}$.

PROOF: By passing to an artinian reduction of A we may assume $\dim(A) = 0$. Then it is clear that A is short with $c = 1$ and $b_h = \dim(A_t) = 1$. Also $(0 : A_1)_{t-1} \neq A_{t-1}$ otherwise $A_t = 0$, hence

$$\beta_h = \dim(0 : A_1)_t + \dim(0 : A_1)_{t-1} < \dim(A_t) + \dim(A_{t-1}) = 1 + \binom{h+t-2}{t-1}.$$

This proves b). The equivalence in c) has been proved in [RV] theorem 3.10. As for d), since $\beta_1 = b_1 + a_1 = b_1 + \binom{h+t-1}{t} - 1$, we get $\beta_1 = \binom{h+t-1}{t}$ if and only if $b_1 = 1$. If $b_1 = 1$, then by b) and c) we get $\beta_h = \binom{h+t-2}{t-1}$. Finally if $\beta_h = \binom{h+t-2}{t-1}$, then by b) and c) we get $b_1 > 0$ and we need only to prove that $\dim(R_{t+1}/R_1 I_t) \leq 1$. Now $\dim(A_t) = 1$ implies $R_t = I_t + kM$ for some monomial M of degree t. Hence we may assume $M = X_1 N$ for some monomial N of degree $t - 1$ and we get

$$R_{t+1} = R_1 I_t + R_1 M = R_1 I_t + X_1 R_1 N \subseteq R_1 I_t + X_1(I_t + kM) = R_1 I_t + kX_1 M$$

This gives the conclusion.

The above Proposition can be applied for example in the following situation.

Corollary 4.3. Let A be a Cohen-Macaulay graded algebra with $e = \binom{h+t-1}{h} + 1$. Let J be the ideal generated by the forms of degree t in I. If $h(J) = h$ then $\beta_1 = \binom{h+t-1}{t} - 1$.

PROOF: As in theorem 4.1 we may assume $\dim(A) = 0$. We have $\dim(I_t) = \binom{h+t-1}{t} - 1$. This implies $R_1 I_t = R_{t+1}$, a fact proved in [RV] theorem 3.10. Hence $b_1 = 0$ and we may apply the above proposition to get the conclusion.

We remark that, again by lemma 3.2, we may apply the above corollary to the case of $e = \binom{t+1}{2} + 1$ points in uniform position in $\mathbf{P}^2$.

The last result of this section gives the Cohen-Macaulay type of some special one-dimensional short graded algebras. This extends a result in [TV].

Theorem 4.4. Let A be a one dimensional short graded algebra with $t = 2$. If $I \subseteq (X_i X_j)_{1 \leq i < j \leq h+1}$ and $X_i X_j \notin I$ for every $i \neq j$, then $\beta_h = b_h = c$.

PROOF: We need only to prove that $a_h = \dim(\operatorname{Tor}_h^R(A, k)_{h+1}) = 0$. The crucial point is that one can compute $\operatorname{Tor}_i^R(A, k)$ via the Koszul resolution of $k = R/(X_1, \ldots, X_{h+1})$

$$0 \to \overset{h+1}{\Lambda} V \otimes R(-h-1) \xrightarrow{\delta_{h+1}} \overset{h}{\Lambda} V \otimes R(-h) \to \cdots \to \Lambda V \otimes R(-1) \xrightarrow{\delta_1} R \to k \to 0$$

where V is a k-vector space of dimension $h+1$. Hence, in order to prove $\operatorname{Tor}_h^R(A,k)_{h+1} = 0$, we need only to prove that the Koszul-type complex

$$\overset{h+1}{\Lambda} V \otimes A(-h-1)_{h+1} \to \overset{h}{\Lambda} V \otimes A(-h)_{h+1} \to \overset{h-1}{\Lambda} V \otimes A(-h+1)_{h+1}$$

is exact in the middle term. We may write this complex in the following way

$$\overset{h+1}{\Lambda} V \otimes k \overset{f=\delta_{h+1}}{\longrightarrow} \overset{h}{\Lambda} V \otimes R_1 \overset{g}{\longrightarrow} \overset{h-1}{\Lambda} V \otimes A_2$$

Now let $\xi \in Ker(g)$; this means that $\delta_h(\xi) \in \overset{h+1}{\Lambda} V \otimes I_2$ and we need to prove that $\xi \in Im(f) = Im(\delta_{h+1}) = Ker(\delta_h)$. This is equivalent to prove that if $\alpha \in \overset{h-1}{\Lambda} V \otimes I_2$ and $\alpha \in Im(\delta_h) = Ker(\delta_{h-1})$, then $\alpha = 0$. Let $e_1, \dots, e_{h+1}$ be a k-vector base of V and $\varepsilon_{ij} = e_1 \wedge \cdots \wedge \hat{e}_i \wedge \cdots \wedge \hat{e}_j \wedge \cdots \wedge e_{h+1}$ be the corresponding vector base of $\overset{h-1}{\Lambda} V$. Then we can write $\alpha = \sum_{1 \le i < j \le h+1} \varepsilon_{ij} \otimes F_{ij}$ with $F_{ij} \in I_2$ and $\delta_{h-1}(\alpha) = 0$. This implies $F_{ij} = \lambda_{ij} X_i X_j$, otherwise if for example $F_{ij} = X_t X_s + \dots$ with $t \ne i,j$ then in $\delta_{h-1}(\alpha)$ we have a term

$$\pm e_1 \wedge \cdots \wedge \hat{e}_i \wedge \cdots \wedge \hat{e}_j \wedge \cdots \wedge \hat{e}_t \wedge \cdots \wedge e_{h+1} \otimes X_t^2 X_s$$

which cannot cancel out since every quadratic form in I_2 does not contain any pure square. This implies that $F_{ij} = 0$ and the conclusion follows.

Corollary 4.5.. *Let A be a one-dimensional short graded algebra with $e = h+2$. If $I \subseteq (X_i X_j)_{1 \le i < j \le h+1}$ and $X_i X_j \notin I$ for every $i \ne j$, then A is Gorenstein.*

We remark that the conditions in the above theorem are verified for a set of $h+1 < e < \binom{h+2}{2}$ points in generic position in $\mathbf{P}^h$ such that $h+1$ of these points are not contained in an hyperplane. On the other hand it is easy to find a short graded algebra with $e = h+2$ which is not Gorenstein.

Let $A = k[X,Y,Z]/(XZ, YZ, X^2Y - XY^2)$; then $h = 2$, $e = 4$, $I \subseteq (XY, XZ, YZ)$ but A is not Gorenstein since it is not a complete intersection.

A remark on a conjecture by Sally

Given a local Cohen-Macaulay ring (A,m) of dimension d, codimension h and multiplicity $e = h+2$, the tangent cone $\operatorname{gr}_m(A) = \oplus m^n/m^{n+1}$ is not necessarily Cohen-Macaulay. But Sally conjectured in [S] that in this case we always have $\operatorname{depth}(\operatorname{gr}_m(A)) \ge d-1$. In the same paper she proves that if $d = 1$, then $H_A(n) \ge h+1$, for every n, hence the Hilbert function of A does not decrease. This implies that $P_A(z) = \frac{1+hz+z^s}{1-z}$ for some $s \ge 2$. Hence we are led to consider graded algebra A, not necessarily Cohen-Macaulay, with Poincare series $P_A(z) = \left(\sum_{i=0}^{t-1} \binom{h+i-1}{i} z^i + z^s \right) / (1-z)^d$ for some integer $s \ge t$. This could be the right notion of short graded algebras in the non Cohen-Macaulay case.

Here we ask the following question. If (A,m) is a Cohen-Macaulay local ring of dimension d, codimension h and multiplicity $e = \binom{h+t-1}{h} + 1$ is it true that $P_A(z) = \left(\sum_{i=0}^{t-1} \binom{h+i-1}{i} z^i + z^s \right) / (1-z)^d$ for some integer s ?

At the moment we are not able to answer this question, but in the case $t = 2$ we can show that this is equivalent to Sally's conjecture.

Proposition 5.1. *Let (A, m) be a local Cohen-Macaulay ring of dimension d, codimension h and multiplicity $e = h + 2$. The following conditions are equivalent.*

a) $\operatorname{depth}(\operatorname{gr}_m(A)) \geq d - 1$.

b) $P_A(z) = \frac{1 + hz + z^s}{(1-z)^d}$.

PROOF: By the result of Sally the conclusion holds in the case $d = 1$. Let $d \geq 2$ and $\operatorname{depth}(\operatorname{gr}_m(A)) \geq d - 1$. We may assume that A/m is infinite and take $x_1, \ldots, x_d$ a minimal reduction of m with x_i superficial for every i. The initial forms $x_1^*, \ldots, x_d^*$ in $\operatorname{gr}_m(A)_1$ are a system of parameters in $\operatorname{gr}_m(A)$, hence we may assume $x_1^*, \ldots, x_{d-1}^*$ form a regular sequence in $\operatorname{gr}_m(A)$. This implies that if $B = A/(X_1, \ldots, X_{d-1})$, then B is a 1-dimensional Cohen-Macaulay ring with the same codimension and multiplicity as A. Further we have $P_A(z) = P_B(z)/(1-z)^{d-1}$. By the result of Sally we get $P_B(z) = \frac{1 + hz + z^s}{1-z}$ for some integer $s \geq 2$ and the conclusion follows. Conversely let us assume $P_A(z) = \frac{1 + hz + z^s}{(1-z)^d}$ and let $B = A/(x_1, \ldots, x_{d-1})$. As before B is a 1-dimensional Cohen-Macaulay ring with the same codimension and multiplicity as A. Since $d \geq 2$ we get $e_1(A) = e_1(B)$, where for a local ring S of dimension d and Poincare series $P_S(z) = \sum_{i=0}^{s} a_i z^i / (1-z)^d$, we define $e_1(S) = \sum_{j=1}^{s} j a_j$ (see [EV]). By the result of Sally we have $P_B(z) = \frac{1 + hz + z^t}{1-z}$, hence $e_1(B) = h + t = e_1(A) = h + s$. This implies $s = t$ and $P_A(z) = P_B(z)/(1-z)^{d-1}$. Hence $x_1^*, \ldots, x_{d-1}^*$ is a regular sequence in $\operatorname{gr}_m(A)$ and the conclusion follows.

Some of the results here were discovered or confirmed with the help of the computer algebra program COCOA written by A.Giovini and G.Niesi.

References

[B] E.BALLICO, *Generators for the homogeneous ideal of s general points in* $\mathbf{P}^3$, J.Alg. **106** (1987), 46–52.

[BK] W.C.BROWN AND J.W.KERR, *Derivations and the Cohen-Macaulay type of points in generic position in n-space*, J.Alg. **112** (1988), 159–172.

[Br1] W.C.BROWN, *A note on the Cohen-Macaulay type of lines in uniform position in* $\mathbf{A}^{n+1}$, Proc. Am. Math. Soc. **87** (1983), 591–595.

[Br2] W.C.BROWN, *A note on pure resolution of points in generic position in* $\mathbf{P}^n$, Rocky Mountain J. Math.. **17** (1987), 479–490.

[EV] J. ELIAS AND G. VALLA, *Rigid Hilbert functions*, J. Pure Appl. Alg. (to appear).

[G] A.V. GERAMITA, *Remarks on the number of generators of some homogeneous ideals*, Bull. Sci. Math., 2x Serie **107** (1983), 193–207.

[GGR] A.V.GERAMITA, D.GREGORY AND L.ROBERTS, *Monomial ideals and points in the projective space*, J.Pure Appl.Alg. **40** (1986), 33–62.

[GO1] A.V.GERAMITA AND F.ORECCHIA, *On the Cohen-Macaulay type of s lines in* $\mathbf{A}^{n+1}$, J.Alg. **70** (1981), 116–140.

[GO2] A.V.GERAMITA AND F.ORECCHIA, *Minimally generating ideals defining certain tangent cones*, J.Alg. **78** (1982), 36–57.

[GM] A.V.GERAMITA AND P.MAROSCIA, *The ideals of forms vanishing at a finite set of points in* $\mathbf{P}^n$, J.Alg. **90** (1984), 528–555.

[HK] J.HERZOG AND M.KUHL, *On the Bettinumbers of finite pure and linear resolution*, Comm.Al. **12** (1984), 1627–1646.

[HM] C.HUNEKE AND M.MILLER, *A note on the multiplicity of Cohen-Macaulay algebras with pure resolutions*, Can. J. Math. **37** (1985), 1149–1162.

[L1] A.LORENZINI, *On the Betti numbers of points in the projective space*, "Thesis," Queen's University, Kingston, Ontario, 1987.

[L2] A.LORENZINI, *Betti numbers of perfect homogeneous ideals*, J.Pure Appl.Alg. **60** (1989), 273–288.

[O] F.ORECCHIA, *Generalised Hilbert functions of Cohen-Macaulay varieties*, "Algebraic Geometry-Open problems,Ravello," Lect. Notes in Math., Springer, 1980, pp. 376–390.

[R] L.G.ROBERTS, *A conjecture on Cohen-Macaulay type*, C.R. Math. Rep. Acad. Sci. Canada **3** (1981), 43–48.

[RV] M.E.ROSSI AND G.VALLA, *Multiplicity and t-isomultiple ideals*, Nagoya Math. J. **110** (1988), 81–111.

[S] J.SALLY, *Cohen-Macaulay local rings of embedding dimension e+d-2*, J.Alg. **83** (1983), 393–408.

[Sc] P.SCHENZEL, *Uber die freien auflosungen extremaler Cohen-Macaulay-ringe*, J.Alg. **64** (1980), 93–101.

[TV] N.V.TRUNG AND G.VALLA, *The Cohen-Macaulay type of points in generic position*, J.Alg. **125** (1989), 110–119.

[W] J.M.WAHL, *Equations defining rational singularities*, Ann. Sci. Ecole Norm. Sup. **10** (1977), 231–264.

Dipartimento di Matematica, Università di Genova, Via L.B.Alberti 4, 16132 Genova, Italy

A Homological approach to symbolic powers

Jürgen Herzog[*]

Introduction

These notes reflect the contents or a lecture given during this workshop in Salvador, August 1988. The intention of this lecture was to introduce some homological methods to study the symbolic powers $I^{(n)}$ of a height two prime ideal I in a three dimensional regular local ring (R, m).

Unfortunately our methods apply only in very special cases, which will be described below. A general theory of the symbolic powers of such ideals is far of being existing. For instance, up to now, one doesn't even know in general the minimal number of generators of $I^{(2)}$. Nevertheless there has been, starting with the paper of Huneke [3], remarkable progress on the question of when the symbolic power algebra $S(I) = \bigoplus_{n>0} I^{(n)}t^n$ is noetherian, see [2,3,4,5,6,7,9]. It is known since the paper of Roberts [8] that $S(I)$ need not always be noetherian. However there is still no non-noetherian symbolic power algebra $S(I)$ known when I is the defining ideal of a monomomial space curve.

In this paper we are mainly interested in the module $I^{(n)}/I^n$, which is a module of finite length and which measures the difference between the n-th symbolic power and the n-th power of I. Huneke [4] has shown that $I^{(n)}/I^n \neq 0$ for $n \geq 2$, if I is not a complete intersection. It is therefore natural to ask the following questions:

1. What is the length of $I^{(n)}/I^n$?
2. What is the number of generators of $I^{(n)}/I^n$?
3. What is the annihilator of $I^{(n)}/I^n$, and how does the exponent $a(n)$ for which $m^{a(n)} \cdot (I^{(n)}/I^n) = 0$ depend on n?

In the first section we collect a few facts, mostly about $I^{(2)}/I^2$, which are more less known. We first show that $(I^{(n)}/I^n)' = \operatorname{Ext}_R^3(I^{(n)}/I^n, R)$ can be presented as the homology of a certain complex. Such a description of $(I^{(n)}/I^n)'$ was first given by Huneke [4], Prop.2.9 in the case that I is generated by three elements. Notice that $(I^{(n)}/I^n)'$ and $I^{(n)}/I^n$ have the same length and the same annihilator.

For $n = 2$, this homological presentation yields the isomorphism $I^{(2)}/I^2 \simeq (\bigwedge^2 \omega)'$, where ω is the canonical module of R/I. It follows immediately from this isomorphism that the ideals $\omega \cdot \omega^{-1}$ and $I_{m-2}(\varphi)$ annihilate the module $I^{(2)}/I^2$. Here m is the minimal number of elements of I, and $I_k(\varphi)$ denotes the ideal generated by the minors of order k of the matrix associated with φ, where

$$0 \to G \xrightarrow{\varphi} F \to I \to 0$$

is the minimal free R-resolution of I. Unfortunately we do not know the exact annihilator of $I^{(2)}/I^2$. If I is an almost complete intersection it has been shown in [4] and [10] that $\operatorname{Ann}(I^{(2)}/I^2) = I_1(\varphi)$, and that $I_1(\varphi) = \omega \cdot \omega^{-1}$ if the number of generators $m = 3$, see [2].

---
[*]Partially supported by DFG and GMD

In the later sections of these notes we assume that I is generated by three elements, so that $\text{rank} F = 3$ and $\text{rank} G = 2$. Let us say that I satisfies the condition n if we can choose bases e_1, e_2, e_3 of F and g_1, g_2 of G such that the matrix of φ with respect to these bases has the form

$$\begin{pmatrix} x_1 & x_2 & x_3 \\ y_1 & y_2 & y_3 \end{pmatrix}$$

with $(y_1, y_2, y_3) \subseteq (x_1, x_2, x_3)^{n-1}$.

Let $D(F)$ be the graded dual of the symmetric algebra $S(F)$. $D(F)$ is a divided power algebra and, associated with $\mathbf{x} = x_1, x_2, x_3$, there is a unique derivation $\partial_1 : D(F) \to D(F)$ with $\partial_1 f_i = x_i$ for $i = 1, 2, 3$. Here f_i is the element in $D_1(F) = F^*$ with $f_i(e_j) = \delta_{ij}$. ∂_1 is a homogeneous R-linear map of degree -1, and hence $\text{Coker} \partial_1$ is a graded R-module. We show in 3.1 that $I^{(k)}/I^k \simeq (\text{Coker} \partial_1)_k$ for $k = 2, \ldots, n$, if I satisfies the condition n.

In section 2 we describe in detail the resolution $C(\mathbf{x})$ of $\text{Coker} \partial_1$. Using this resolution we able to answer the above questions on $I^{(k)}/I^k$ for $k = 2, \ldots, n$: We show that $I^{(k)}/I^k$ is a self-dual module (3.1), and is minimally generated by $\binom{k}{2}$ elements, see 3.3. We compute the length of $I^{(k)}/I^k$ and show in Corollary 3.4 that $S(I)$ needs at least $3 + \binom{n+1}{3}$ generators over R, and compute explicitly the three generators of $I^{(3)}/I^3$ in the monomial case, see 3.6. Finally we describe in 3.7 the minimal R-free resolution of $I^{(k)}/I^k$.

I wish to thank Winfried Bruns, Bernd Ulrich and Wolmer Vasconcelos for many helpful discussions and clarifying comments on this topic. In particular I am indebted to Wolmer Vasconcelos who invited me in Spring 1988 to visit Rutgers University where I started this work, and to Aron Simis who invited me to participate the workshop in Salvador where I had the occasion to discuss and to present this material.

1 Preliminaries

Let (R, m) be a three-dimensional regular local ring, and let $I \subseteq R$ be a prime ideal of height two. In this section we want to describe the module $\text{Ext}_R^3(I^{(n)}/I^n, R)$ (which is the 'dual' of $I^{(n)}/I^n$) as the homology of a certain complex and to draw some simple consequences from this presentation.

Let us first recall a few facts from multilinear algebra which will be used here and in later sections: Let R be an arbitrary commutative ring (with 1) and let F be a free R-module of finite rank. We denote by $S(F)$ the symmetric algebra of F. $S(F)$ is a graded R-algebra whose i-th homogeneous component $S_i(F)$ is the i-th symmetric power of F. We denote by $D(F)$ the graded dual of $S(F)$, that is, $D(F) = \bigoplus_{n \geq 0} S_n(F)^*$. Here M^* denotes the R-dual of an R-module M.

$D(F)$ has a natural structure of a divided power algebra. Let $\varphi \in S_i(F)^*, \psi \in S_j(F)^*$ and $n = i + j$, then $\varphi \cdot \psi \in S_n(F)^*$ is easy to describe on decomposable elements. Let $a_1, \ldots, a_n \in S_1(F) = F$. For any subset I of $J = \{1, \ldots, n\}$, we set $a_I = \prod_{k \in I} a_k$. With this notation we have $(\varphi \cdot \psi)(a_J) = \sum_{I \subseteq J, |I| = i} \varphi(a_I) \cdot \psi(a_{J \setminus I})$. Moreover, if $\varphi \in S_1(F)^*$, then the n-th symbolic power of φ is defined by the equation $\varphi^{(n)}(a_J) = \prod_{i=1}^n \varphi(a_i)$. It is clear from these definitions that $\varphi^n = n! \varphi^{(n)}$, and so φ^n may be zero in positive characteristic. If $\varphi_1, \ldots, \varphi_m$ is a basis of F^*, then $\{\varphi_1^{(i_1)} \cdot \ldots \cdot \varphi_m^{(i_m)} \mid \sum_{j=1}^m i_j = n\}$ is a basis of $S_n(F)^*$.

Now we come back to our ideal I. Suppose I is minimally generated by m elements, then I has a minimal R-free resolution

$$0 \to G \xrightarrow{\varphi} F \to I \to 0$$

where $\mathrm{rank}\, G = \mathrm{rank}\, F - 1 = m - 1$.

Weyman's [11] 'n-th symmetric power' of the complex $K. : 0 \to G \xrightarrow{\varphi} F \to 0$, which in this case coincides with the so-called Z-complex (see [1]), has the form

$$S_n(K.) : \cdots \xrightarrow{\partial_3} \overset{2}{\bigwedge} G \otimes S_{n-2}(F) \xrightarrow{\partial_2} G \otimes S_{n-1}(F) \xrightarrow{\partial_1} S_n(F) \to 0$$

The 0-th homology of this complex is isomorphic to the n-th symmetric power $S_n(I)$ of I. Therefore there is a natural map $\partial_0 : S_n(F) \to I^n$, and we obtain the augmented complex

$$\widetilde{S_n(K.)} : \cdots \to \overset{2}{\bigwedge} G \otimes S_{n-2}(F) \xrightarrow{\partial_2} G \otimes S_{n-1}(F) \xrightarrow{\partial_1} S_n(F) \xrightarrow{\partial_0} I^n \to 0$$

This complex has homology of finite length, and it is exact if and only if I is generated by at most three elements. Nevertheless we may use this complex to compute $\mathrm{Ext}_R^i(I^n, R)$ for $i = 0, 1, 2$. The following elementary lemma explains why.

Lemma 1.1 *Let* (R, m) *be a local Cohen-Macaulay ring of dimension d, and let*

$$\tilde{F}. : \cdots \to F_2 \xrightarrow{\partial_2} F_1 \xrightarrow{\partial_1} F_0 \xrightarrow{\partial_0} F_{-1} \to 0$$

be a complex such that
  1) F_i is free for $i > 0$
  2) $H_{-1}(\tilde{F}.) = 0$, and $H_i(\tilde{F}.)$ has finite length for $i \geq 0$.

Then

$$\mathrm{Ext}_R^i(F_{-1}, R) \simeq H^i(F^*.) \ for \ i = 0, 1, \cdots, d - 1,$$

where $F^*.$ is the complex $0 \to F_0^* \xrightarrow{\partial_0^*} F_1^* \xrightarrow{\partial_1^*} \cdots$

PROOF. Set $Z_i = \mathrm{Ker}\, \partial_i$, $B_i = \mathrm{Im}\, \partial_i$ and $Z_{-1} = F_{-1}$. If M is a module of finite length then $\mathrm{Ext}_R^j(M, R) = 0$ for $j = 0, \ldots, d - 1$. Therefore the exact sequences

$$0 \to B_i \to Z_i \to H_i(\tilde{F}.) \to 0$$

induce isomorphisms

$$\mathrm{Ext}_R^j(Z_i, R) \simeq \mathrm{Ext}_R^j(B_i, R)$$

for all i and for all $j = 0, \ldots, d - 2$.

Using these isomorphisms one proves by induction on j that

$$\mathrm{Ext}_R^j(F_{-1}, R) \simeq \mathrm{Ext}_R^1(B_{j-2}, R)$$

for $2 \leq j \leq d - 1$.

Now, since $Z^*_{j-1} = B^*_{j-1}$ for all j, the exact sequences

$$0 \to Z_{j-1} \to F_{j-1} \to B_{j-2} \to 0$$

yield the exact sequences

$$0 \to B^*_{j-2} \to F^*_{j-1} \to B^*_{j-1} \to \operatorname{Ext}^1_R(B_{j-2}, R) \to 0$$

For $j = 1$, this exact sequence implies that $\operatorname{Ext}^i_R(F_{-1}, R) \simeq H^i(F^*)$ for $i = 0, 1$, and for $j \geq 2$, it implies that $\operatorname{Ext}^1_R(B_{j-2}, R) \simeq H^j(F^*)$, as we wanted to show. $\square$

Let M be an R-module of finite length. We set $M' = \operatorname{Ext}^3_R(M, R)$. Notice that $\operatorname{length} M = \operatorname{length} M'$, and that $M \simeq M''$.

Let us denote by ω_n the canonical module of $R/I^{(n)}$. If we apply the previous lemma to the complex $S_n(K.)$ we obtain

Lemma 1.2 *The complex*

$$S_n(K.)^* : 0 \to D_n(F) \xrightarrow{\partial^*_1} G^* \otimes D_{n-1}(F) \xrightarrow{\partial^*_2} \overset{2}{\bigwedge} G^* \otimes D_{n-2}(F) \xrightarrow{\partial^*_3} \overset{3}{\bigwedge} G^* \otimes D_{n-3}(F) \to$$

has the following homology

$$H^0(S_n(K.)^*) \simeq R,$$
$$H^1(S_n(K.)^*) \simeq \omega_n,$$
$$H^2(S_n(K.)^*) \simeq (I^{(n)}/I^n)'.$$

PROOF. The exact sequence

$$0 \to I^n \to I^{(n)} \to I^{(n)}/I^n \to 0$$

gives rise to isomorphisms

$$\operatorname{Ext}^1_R(I^n, R) \simeq \operatorname{Ext}^1_R(I^{(n)}, R) \simeq \operatorname{Ext}^2_R(R/I^{(n)}, R) \simeq \omega_n$$

and

$$\operatorname{Ext}^2_R(I^n, R) \simeq \operatorname{Ext}^3_R(R/I^n, R) \simeq (I^{(n)}/I^n)'.$$

$\square$

We must admit that the usefulness of this lemma is limited since we have no means to actually compute the homology.

There are however two instances where $H^2(S_n(K.)^*)$ is simply isomorphic to $\operatorname{Coker}\partial^*_2$. This has been first noted by C. Huneke in [3], Prop.2.9.

Corollary 1.3 $(I^{(n)}/I^n)' \simeq \operatorname{Coker}(G^* \otimes D_{n-1}(F) \xrightarrow{\partial^*_2} \bigwedge^2 G^* \otimes D_{n-2}(F))$ *if $n = 2$, or if I is generated by 3 elements.*

To simplify notations we write in the sequel ω instead of ω_1

Corollary 1.4 $I^{(2)}/I^2 \simeq (\bigwedge^2 \omega)'$

PROOF. $\omega \simeq \text{Ext}^1_R(R/I, R) \simeq \text{Coker}(F^* \xrightarrow{\varphi^*} G^*)$, and $(I^{(2)}/I^2)' \simeq \text{Coker}(G^* \otimes F^* \to \bigwedge^2 G^*) \simeq \bigwedge^2 \omega$. $\square$

We may identify ω with an ideal in R/I. It is then clear that $\omega \cdot \omega^{-1}$ annihilates $\bigwedge^2 \omega$, and so we get

Corollary 1.5 $(\omega \cdot \omega^{-1}) \cdot (I^{(2)}/I^2) = 0$

We denote by $I_j(\varphi)$ the ideal generated by the minors of order j of the matrix associated with $\varphi : G \to F$ with respect to some bases of F and G.

Corollary 1.6 $I_{m-2}(\varphi) \cdot (I^{(2)}/I^2) = 0$

PROOF. We show that $I_{m-2}(\varphi^*) \cdot R/I \subseteq \omega \cdot \omega^{-1}$. In fact, there is an exact sequence

$$(R/I)^m \xrightarrow{\overline{A}} (R/I)^{m-1} \to \omega \to 0$$

where $A = (a_{ij})$ is the matrix associated with φ^*, and where $\overline{A} = (\overline{a}_{ij})$ is the matrix whose entries $\overline{a}_{ij}$ are the residue classes of the a_{ij} modulo I. Pick any $m-2$ rows of $\overline{A}$ and call the matrix with these rows B. Let Δ_i be the maximal minor of B which is obtained from B by deleting the i-th column, and let $\Delta = (\Delta_1, -\Delta_2, \ldots, (-1)^{m-1}\Delta_{m-1})^t$. Then $B \cdot \Delta = 0$, and so $\overline{A} \cdot \Delta = 0$ since $\text{rank}\overline{A} = m - 2$. This implies that there is an (R/I)-module homomorphism $\alpha_\Delta : \omega \to R$ with $\text{Im}\alpha_\Delta = (\Delta_1, \ldots, \Delta_{m-1})$. Since $\omega \cdot \omega^{-1} = \sum_\alpha \alpha(\omega)$, where the sum is taken over all homomorphisms $\alpha : \omega \to R$, we conclude that $(\Delta_1, \ldots, \Delta_{m-1}) \subseteq \omega \cdot \omega_{-1}$. As we have chosen the $m-2$ rows of $\overline{A}$ arbitrarily, the conclusion follows. $\square$

Note that $I_1(\varphi)$ is exactly the annihilator of $I^{(2)}/I^2$, if I is an almost complete intersection, as has been observed by C.Huneke [4] and Vasconcelos [9].

We conclude this section by giving a rough estimate on the length of $I^{(2)}/I^2$.

Corollary 1.7 $\text{length}(I^{(2)}/I^2) \geq \binom{m-1}{2} \cdot \text{length}(R/I_:(\varphi))$

PROOF. It is clear that $\text{Im}\partial_2^*$ is a submodule of $I_1(\varphi) \cdot (\bigwedge^2 G^* \otimes D_{n-2}(F))$. Hence the assertion follows from 1.3. $\square$

2   The complex $C(n, \mathbf{x})$

In this section we construct complexes, denoted $C(n, \mathbf{x})$, which will be used later to study symbolic powers of certain ideals. $C(n, \mathbf{x})$ will be the $(n-2)$-th homogeneous component of a complex $C(\mathbf{x})$ which we associate with a sequence $\mathbf{x} = x_1, x_2, x_3$. Thus we will have $C(\mathbf{x}) = \bigoplus_n C(n, \mathbf{x})$.

To introduce $C(\mathbf{x})$ we let F be a free module of finite rank and choose an element $\mathbf{x} \in F$. Then $\mathbf{x}$ induces a homogeneous R-linear map $\mu_{\mathbf{x}} : S(F)(-1) \to S(F)$, with $\mu_{\mathbf{x}}(a) = \mathbf{x} \cdot a$ for all $a \in S(F)$, and hence induces the dual homogeneous R-linear map $\mu_{\mathbf{x}}^* : D(F)(+1) \to D(F)$.

Lemma 2.1 $\mu_{\mathbf{x}}^*$ *is a derivation, which means that* $\mu_{\mathbf{x}}^*$ *satisfies*
  a) $\mu_{\mathbf{x}}^*(\varphi \cdot \psi) = \varphi \cdot \mu_{\mathbf{x}}^*(\psi) + \psi \cdot \mu_{\mathbf{x}}^*(\varphi)$ *and*
  b) $\mu_{\mathbf{x}}^*(\varphi^{(n)}) = \varphi^{(n-1)} \cdot \mu_{\mathbf{x}}^*(\varphi)$ *for* $\varphi \in F^*$ *and all* n.

PROOF. Let $e_1, \ldots, e_m$ be a basis of F and let $f_1, \ldots, f_m$ be the basis of F^* which is dual to $e_1, \ldots, e_m$. If $\mathbf{x} = \sum_{i=1}^{m} x_i e_i$, then it is easy to verify that

$$\mu_{\mathbf{x}}^*(f_1^{(i_1)} \cdot \ldots \cdot f_m^{(i_m)}) = \sum_{j=1}^{m} x_j f_1^{(i_1)} \cdot \ldots \cdot f_j^{(i_j-1)} \cdot \ldots \cdot f_m^{(i_m)}$$

Here we use the convention that $f^{(a)} = 0$ for $a < 0$. This equation implies immediately a) and b). $\square$

Remark 2.2 Notice that a derivation $\partial : D(F)(-1) \to D(F)$ is uniquely determined by its restriction to $D_1(F) = F^*$. Using the notations of the proof of 2.1 we have in our particular case: $\mu_{\mathbf{x}}^*(f_i) = x_i$ for $i = 1, \ldots, n$.

We need one more observation from multilinear algebra. Let G be another free R-module of finite rank, and let $\Phi : S(G) \to D(F)$ be a homogeneous algebra homomorphism.

Lemma 2.3 *The dual map $\Phi^* : S(F) \to D(G)$ is again a homogeneous R-algebra homomorphism.*

PROOF. Let $a \in S_n(F)$ and $b \in S_n(G)$ then $\Phi^*(a)$ is the element of $D_n(G) = S_n(G)^*$ for which $\Phi^*(a)(b) = \Phi(b)(a)$. Let $a = a_1 \cdot \ldots \cdot a_n$ with $a_i \in G$ and $b = b_1 \cdot \ldots \cdot b_n$ with $b_i \in F$ then

$$\Phi^*(a)(b) = \sum_{\pi \in S_n} \prod_{i=1}^{n} \Phi(b_i)(a_{\pi(i)})$$

$$= \sum_{i=1}^{n} \Phi(b_i)(a_1) \cdot \Phi(b_2 \cdot \ldots \widehat{b_i} \ldots \cdot b_n)(a_2 \cdot \ldots \cdot a_n)$$

$$= \sum_{i=1}^{n} \Phi^*(a_1)(b_i) \cdot \Phi^*(a_2 \cdot \ldots \cdot a_n)(b_2 \cdot \ldots \widehat{b_i} \ldots \cdot b_n) = \Phi^*(a_1) \cdot \Phi^*(a_2 \cdot \ldots \cdot a_n)(b)$$

These equations imply that Φ^* is a homomorphism. $\square$

We are now ready to define the complex $C(\mathbf{x})$. Let F be a free R-module with basis e_1, e_2, e_3 and F^* the dual R-module with dual basis f_1, f_2, f_3. Associated with $\mathbf{x} = x_1 e_1 + x_2 e_2 + x_3 e_3$ we define $C(\mathbf{x})$ to be

$$0 \to S(F) \xrightarrow{\partial_3} S(F)(+1) \xrightarrow{\partial_2} D(F)(+1) \xrightarrow{\partial_1} D(F) \to 0$$

where the homomorphisms ∂_i are defined as follows: $\partial_3 = \mu_{\mathbf{x}}$, $\partial_1 = \mu_{\mathbf{x}}^*$, while ∂_2 is the homomorphism of graded R-algebras given by

$$\partial_2(e_1) = x_2 f_3 - x_3 f_2$$

$$\partial_2(e_2) = x_3 f_1 - x_1 f_3$$

$$\partial_2(e_3) = x_1 f_2 - x_2 f_1$$

Proposition 2.4 a) $C(\mathbf{x})$ *is a self-dual homogeneous complex of graded R-modules. (R is equipped with the trivial grading)*

b) *If R contains the rational numbers Q and* $\mathbf{x} = x_1, x_2, x_3$ *is a regular sequence then* $C(\mathbf{x})$ *is exact.*

PROOF. a) ∂_2 is an algebra homomorphism, and since $\mathbf{x} \in \text{Ker}\partial_2$, it follows that $\text{Im}\partial_3 = \mathbf{x} \cdot S(F) \subseteq \text{Ker}\partial_2$, and hence $\partial_3 \circ \partial_2 = 0$. Dualizing we obtain $\partial_2^* \circ \partial_3^* = \partial_2^* \circ \partial_1$. But $\partial_2^*|_{S_i(F)} = -\partial_2|_{S_i(F)}$, and since ∂_2^* is an R-algebra homomorphism (see 2.3), we conclude that $\partial_2|_{S_i(F)} = (-1)^i \partial_2^*|_{S_i(F)}$ for all i, and so $\partial_2 \circ \partial_1 = 0$. These computations also show that the complexes $C(\mathbf{x})$ and $C(\mathbf{x})^*$ are isomorphic, and hence self-dual.

b) Assume that $H_i(C(\mathbf{x})) \neq 0$ for some $i > 0$. After localization at a prime ideal we may assume that $H_i(C(\mathbf{x}))$ has finite length for all $i > 0$.

If $(x_1, x_2, x_3) \neq R$ then depth$C_i(\mathbf{x}) \geq 3$ for all i, and the acyclicity lemma implies that $C(\mathbf{x})$ is acyclic.

If $(x_1, x_2, x_3) = R$, then one of the x_i, say x_1, must be a unit. We may as well assume that $x_1 = 1$, and choose the new basis

$$
\begin{aligned}
e_1' &= e_1 + x_2 e_2 + x_3 e_3 \\
e_2' &= e_2 \\
e_3' &= e_3
\end{aligned}
$$

Let f_i' be the dual element of e_i' in F^* then

$$
\begin{aligned}
f_1' &= f_1 \\
f_2' &= -x_2 f_1 + f_2 \\
f_3' &= -x_3 f_1 + f_3
\end{aligned}
$$

and $\partial_3(1) = e_1'$, $\partial_2(e_1') = 0$, $\partial_2(e_2') = -f_3'$, $\partial_2(e_3') = -f_2'$. Finally, ∂_1 is the derivation with $\partial_1(f_1') = 1$ and $\partial_1(f_i') = 0$ for $i = 2, 3$.

It is then clear that ∂_3 is injective, that $\text{Ker}\partial_2 = e_1' S(F) = \text{Im}\partial_3$, and that ∂_1 is an epimorphism whose kernel is generated over R by the elements $f_2'^{(a)} f_3'^{(b)}$, $a, b \geq 0$. It follows that $\text{Ker}\partial_1 = R[f_2', f_3']$ if $Q \subseteq R$. But this is exactly the image of ∂_2. □

Remark 2.5 Let F be a free module with basis $f_1, \ldots, f_m$ and consider the derivation $\partial : D(F)(+1) \to D(F)$ with $\partial(f_i) = x_i$ for $i = 1, \cdots, m$. Ker∂ is a subring of $D(F)$. Suppose $x_1, \cdots, x_m$ is a regular sequence. One readily verifies that in this case, as for $m = 3$, the subring Ker∂ is generated over R by the elements $x_i e_j - x_j e_i$ for all i, j with $1 \leq i < j \leq m$. Thus Ker∂ is an epimorphism of a polynomial ring $T = R[\{T_{ij}\}_{1 \leq i < j \leq m}]$. The homogeneous free T-resolution of Ker∂ composed with ∂ then yields free R-resolutions of the homogeneous components of Coker∂. In case $m = 3$ we had $T = S(F)$ and Ker$\partial = T/(\mathbf{x})T$.

Definition 2.6 Let $n \geq 2$ be an integer. Then $C(n, \mathbf{x})$ is the $(n-2)$-th homogeneous component of $C(\mathbf{x})$

Hence, if we let S_i (resp. D_i) denote the i-th homogeneous component of $S(F)$ (resp. $D(F)$) then

$$
C(n, \mathbf{x}) : 0 \to S_{n-2} \xrightarrow{\partial_3} S_{n-1} \xrightarrow{\partial_2} D_{n-1} \xrightarrow{\partial_1} D_{n-2} \to 0
$$

By 2.4 these complexes are self-dual, and they are acyclic provided $Q \subseteq R$ and $\mathbf{x}$ is a regular sequence.

Before we go on, we show by an example that the assumption $Q \subseteq R$ is essential. Let $x_1, x_2, x_3 \in R$ be a regular sequence and assume char$R = 2$. Then Coker∂_1, with ∂_1 as in $C(n, \mathbf{x})$, has the following resolution:

$$0 \to R \xrightarrow{\partial_3} \bigoplus_{i=0}^{3} R \cdot a_i \xrightarrow{\partial_2} D_2 \xrightarrow{\partial_1} D_1 \to 0$$

with

$$
\begin{aligned}
\partial_2(a_0) &= x_1 f_2 f_3 + x_2 f_1 f_3 + x_3 f_1 f_2 \\
\partial_2(a_1) &= x_2 x_3 f_2 f_3 + x_2^2 f_3^{(2)} + x_3^2 f_2^{(2)} \\
\partial_2(a_2) &= x_1 x_3 f_1 f_3 + x_1^2 f_3^{(2)} + x_3^2 f_1^{(2)} \\
\partial_2(a_3) &= x_1 x_2 f_1 f_2 + x_1^2 f_2^{(2)} + x_2^2 f_1^{(1)} \\
\partial_3(1) &= x_1 x_2 x_3 a_0 + x_1^2 a_1 + x_2^2 a_2 + x_3^2 a_3
\end{aligned}
$$

For the rest of this section we will assume that $Q \subseteq R$, and we set

$$H_i(n, \mathbf{x}) := H_i(C(n, \mathbf{x}))$$

Let $R = A[T_1, T_2, T_3]$ be a polynomial ring in the variables T_1, T_2, T_3 over the commutative ring A. We consider R in the natural way as a graded A-algebra. It is then clear that $C(n, T)$ becomes a homogeneous complex

$$0 \to S_{n-2}(-n-1) \to S_{n-1}(-n) \to D_{n-1}(-1) \to D_{n-2} \to 0$$

of graded R-modules. In particular, $H_0(n, T)$ is a graded R-module.

Proposition 2.7 a) $H_0(n, T)_i$ *is a free A-module for all i.*

b)

$$\left(\sum_{i \geq 0} \mathrm{rank}_A H_0(n, T)_i \cdot t^i \right)(1 - t)^3$$

$$= \binom{n}{2} - \binom{n+1}{2}t + \binom{n+1}{2}t^n - \binom{n}{2}t^{n+1}$$

c) $H_0(n, T)_i = 0$ *for $i > n - 2$.*

PROOF. b) follows at once from the homogeneous resolution of $H_0(n, T)$. Now b) implies that $\mathrm{rank}_A H_0(n, T)_i = 0$ for $i > n - 2$, and hence a) implies c).

In order to prove a) we show that $\mathrm{Tor}_i^A(A/I, H_0(n, T)) = 0$ for $i > 0$ and all ideals $I \subseteq A$. In fact, $\mathrm{Tor}_i^A(A/I, H_0(n, T)) = H_i(A/I \otimes_A C_A(n, T)) = 0$, since $A/I \otimes_A C_A(n, T) \simeq C_{A/I}(n, T)$ is acyclic as T_1, T_2, T_3 is a regular sequence in $(A/I)[T_1, T_2, T_3]$. $\square$

Corollary 2.8 *Let $\mathbf{x} = x_1, x_2, x_3$ be a sequence of elements of R. Then*
a) $(x_1, x_2, x_3)^{n-1} \cdot H_0(n, \mathbf{x}) = 0$
b) *If $\mathbf{x}$ is a regular sequence and $c \in H_0(n, \mathbf{x}) \setminus (x_1, x_2, x_3)H_0(n, \mathbf{x})$, then $(x_1, x_2, x_3)^{n-2} \cdot c \neq 0$.*

PROOF. a) follows immediately from 2.7, c) by specializing the T_i to x_i.

b) We set $I = (x_1, x_2, x_3)$ and $H_0 = H_0(n, \mathbf{x})$. The associated graded module $gr_I H_0 = \bigoplus_{j \geq 0} I^j H_0 / I^{j+1} H_0$ is a module over the associated graded ring $B = gr_I R = \bigoplus_{j \geq 0} I^j / I^{j+1}$ which is isomorphic to the polynomial ring $(R/I)[\xi_1, \xi_2, \xi_3]$, where ξ_i is the leading form of x_i for $i = 1, 2, 3$. We first show that $gr_I H_0 \simeq H_0(n, \xi)$. To this end we define the following filtration $\mathcal{F}$ on $C(n, \mathbf{x})$: For all $j \geq 0$ we set $\mathcal{F}_j D_{n-2} = I^j D_{n-2}$, $\mathcal{F}_j D_{n-1} = I^{j-1} D_{n-1}$, $\mathcal{F}_j S_{n-1} = I^{j-n} S_{n-1}$ and $\mathcal{F}_j S_{n-2} = I^{j-n-1} S_{n-2}$. (Of course we let $I^a = R$ for $a < 0$, as usual).

It is immediate that

$$gr_\mathcal{F} C_R(n, \mathbf{x}) = C_B(n, \xi)$$

As ξ is a regular sequence in B we conclude that $C(\xi)$ is acyclic. But this implies that $gr_I H_0(n, \mathbf{x}) \simeq H_0(n, \xi)$.

Now suppose that statement b) is false. Then there exists $c \in H_0(n, \xi)_0$, $c \neq 0$ such that $(\xi_1, \xi_2, \xi_3)^{n-2} \cdot c = 0$, and hence a homogeneous homomorphism

$$\gamma : B/(\xi_1, \xi_2, \xi_3)^{n-2} \to H_0(n, \xi)$$

such that $\gamma(1) = c$. Now γ induces a homogeneous complex homomorphism γ_* between the free B-resolutions of $B/(\xi_1, \xi_2, \xi_3)^{n-2}$ and $H_0(n, \xi)$:

$$
\begin{array}{ccccccccc}
0 & \to & B^{b_3}(-n) & \to & B^{b_2}(-n+1) & \to & B^{b_1}(-n+2) & \to & B^{b_0} & \to & 0 \\
 & & \downarrow \gamma_3 & & \downarrow \gamma_2 & & \downarrow \gamma_1 & & \downarrow \gamma_0 & & \\
0 & \to & S_{n-2}(-n-1) & \to & S_{n-1}(-n) & \to & D_{n-1}(-1) & \to & D_{n-2} & \to & 0
\end{array}
$$

Since γ_3 is homogeneous, it follows from the shifts in the diagram that $\gamma_3 = 0$, and so $\mathrm{Ext}_B^3(\gamma, B) = 0$. As the complexes which are dual to the resolutions of $B/(\xi_1, \xi_2, \xi_3)^{n-2}$ and $H_0(n, \xi)$ are again acyclic, we conclude that $\gamma = \mathrm{Ext}_B^3(\mathrm{Ext}_B^3(\gamma, B), B)$, and hence $\gamma = 0$. This is a contradiction. $\square$

3   Applications to symbolic powers

Let (R, m) be a 3-dimensional regular local ring containing the rational numbers, $I \subseteq R$ a prime ideal of height 2 generated by three elements, and let

$$0 \to G \xrightarrow{\varphi} F \to I \to 0$$

be the minimal free R-resolution of I. We then have $\mathrm{rank} F = 3$ and $\mathrm{rank} G = 2$.

The relevance of the complexes $C(n, \mathbf{x})$ for the symbolic powers of I is given by the following

Theorem 3.1 *Suppose there exist bases $\mathcal{A}$ of F and $\mathcal{B}$ of G such that the matrix*

$$A = \begin{pmatrix} x_1 & x_2 & x_3 \\ y_1 & y_2 & y_3 \end{pmatrix}$$

describing φ with respect to $\mathcal{A}$ and $\mathcal{B}$ satisfies

$$(y_1, y_2, y_3) \subseteq (x_1, x_2, x_3)^{n-1}$$

for some $n \geq 2$. Then for $k = 2, \ldots, n$ one has:

a) $I^{(k)}/I^k$ *is a self-dual module, which means that* $I^{(k)}/I^k \simeq \operatorname{Ext}^3_R(I^{(k)}/I^k, R)$.

b) $I^{(k)}/I^k \simeq H_0(k, \mathbf{x})$.

PROOF. Let $\mathcal{A} = e_1, e_2, e_3$ and $\mathcal{B} = g_1, g_2$. We know from 1.3 that $\operatorname{Ext}^3_R(I^{(k)}/I^k, R) \simeq$ Cokerα, where $\alpha : D_{k-1}(F) \otimes G^* \to D_{k-2}(F)$ is defined by

$$\alpha(f \otimes a_1 g_1 + a_2 g_2) = a_1 \mu_{\mathbf{x}}^*(f) + a_2 \mu_{\mathbf{y}}^*(g)$$

with

$$\mathbf{x} = \varphi(g_1) = x_1 e_1 + x_2 e_2 + x_3 e_3$$
$$\mathbf{y} = \varphi(g_2) = y_1 e_1 + y_2 e_2 + y_3 e_3$$

and where μ^* is defined as in 2.1.

Notice that Coker$(\alpha|_{D_{k-1}(F) \otimes Rg_1}) = H_0(k, \mathbf{x})$. Now as

$$\operatorname{Im}(\alpha|_{D_{k-1}(F) \otimes Rg_2}) \subseteq (y_1, y_2, y_3) D_{k-2}(F),$$

it follows from 2.8, a) and our assumption $(y_1, y_2, y_3) \subseteq (x_1, x_2, x_3)^{n-1}$ that

$$\operatorname{Ext}^3_R(I^{(k)}/I^k, R) \simeq \operatorname{Coker}\alpha = \operatorname{Coker}(\alpha|_{D_{k-1}(F) \otimes Rg_1}) = H_0(k, \mathbf{x})$$

Since the prime ideal I of height two is not a complete intersection, we have grade$I_1(A) \geq 3$. But $I_1(A) = (x_1, x_2, x_3)$, and so x_1, x_2, x_3 is a regular sequence. Thus we see that $C(k, \mathbf{x})$ is an R-free resolution of $\operatorname{Ext}^3_R(I^{(k)}/I^k, R)$. However, since $C(k, \mathbf{x})$ is self-dual we conclude that $C(k, \mathbf{x})$ is a free R-resolution of $I^{(k)}/I^k$ as well. In particular, it follows that $I^{(k)}/I^k$ is self-dual and that $I^{(k)}/I^k = H_0(k, \mathbf{x})$. $\square$

Remark 3.2 The arguments of the proof of 3.1 show actually the following: If I is a height two prime ideal with three generators in a regular local ring, then $(I^{(k)}/I^k)'$ is a factor module of $H_0(k, \mathbf{x})$, where $\mathbf{x}$ can be chosen to be the first or second row of the relation matrix A of I. We know from 2.8, a) that $(\mathbf{x})^{k-1} H_0(k, \mathbf{x}) = 0$, and hence we see that

$$I_1(A)^{2(k-1)}(I^{(k)}/I^k) = 0 \text{ for all } k \geq 0.$$

We use our results from section 2 to obtain some more informations about the structure of $I^{(k)}/I^k$.

Corollary 3.3 *Under the assumptions of 3.1, we have for all $k = 1, \ldots, n$:*

a) $I^{(k)}/I^k$ *is minimally generated by* $\binom{k}{2}$ *elements.*

b) length$(I^{(k)}/I^k) = ((\binom{k+1}{3})(\binom{k}{2}) - (\binom{k+1}{2})(\binom{k}{3}))$length$(R/I_1(A))$

c) $I_1(A)^{k-1}(I^{(k)}/I^k) = 0$

d) *If $c \in I^{(k)} \setminus I_1(A) \cdot I^{(k)}$, then $I_1(A)^{k-2} \cdot c \not\subseteq I^k$.*

PROOF. Let us denote by $\mu(M)$ the minimal number of generators of a module M. Using that $I^{(k)}/I^k \simeq H_0(k, \mathbf{x})$, we get $\mu(I^{(k)}/I^k) = \mu(H_0(k, \mathbf{x})) = \mu(D_{k-2}) = \binom{k}{2}$. This proves a). The assertions c) and d) follow immediately from 2.8 and 3.1, b) .

$B = gr_{I_1(A)}(R)$ is the polynomial ring $(R/I_1(A))[\xi_1, \xi_2, \xi_3]$ and $gr_{I_1(A)}(H_0(k, \mathbf{x})) \simeq H_0(k, \xi)$, see proof of 2.8, a). It is therefore clear that

$$\text{length}(I^{(k)}/I^k) = \text{length}(H_0(k, \mathbf{x})) = \text{length}(H_0(k, \xi))$$

By 2.7, a) each homogeneous component of $H_0(k, \xi)$ is a free $R/I_1(A)$-module, and thus we have $\text{length}(H_0(k, \xi)) = \sum_i \text{rank}(H_0(k, \xi)_i)$.

Let $P(t) = \binom{k}{2} - \binom{k+1}{2}t + \binom{k+1}{2}t^k - \binom{k}{2}t^{k+1}$ and let $P^{(3)}(t)$ denote the third derivative of $P(t)$, then the equation 2.7, b) implies that

$$\sum_i \text{rank}(H_0(k, \xi)_i) = -P^{(3)}(1)$$

$$= \binom{k+1}{3}\binom{k}{2} - \binom{k+1}{2}\binom{k}{2}$$

Corollary 3.4 Let $S = \bigoplus_{k \geq 0} I^{(k)}t^k$ be the symbolic power algebra of the ideal I satisfying the assumptions of 3.1. For $k = 2, \ldots, n$ the algebra S needs exactly $\binom{k}{2}$ generators in degree k, and three generators in degree 1. Therefore the number of generators of S as an R-algebra is at least

$$3 + \sum_{k=2}^{n} \binom{k}{2} = 3 + \binom{n+1}{3}$$

PROOF. Let $k \geq 2$, and set $\Sigma = \sum_{j=1}^{k-1} I^{(j)} I^{(k-j)}$. Then the number of generators needed in degree k equals $\mu(I^{(k)}/\Sigma)$. Since $I^k \subseteq \Sigma$, we get an exact sequence

$$0 \to \Sigma/I^k \xrightarrow{\iota} I^{(k)}/I^k \to I^{(k)}/\Sigma \to 0$$

For $j = 1, \ldots, k-1$ we have $I_1(A)^{j-1} \cdot I^{(j)} \subseteq I^j$, see 2.8, a) and 3.1, b) . Therefore $I_1(A)^{k-2} \cdot I^{(j)} \cdot I^{(k-j)} \subseteq I^j \cdot I^{k-j} \subseteq I^k$ for all $j = 1, \ldots, k$, and hence $I_1(A)^{k-2} \cdot (\Sigma/I^k) = 0$. Now 2.8, b) yields $\text{Im}\iota \subseteq I_1(A) \cdot (I^{(k)}/I^k)$, and the assertion follows. $\square$

The next proposition tells us how to compute the generators of $I^{(k)}/I^k$ for those k which satisfy the conditions of 3.1. Dualizing the resolution

$$S(k) : 0 \to S_{k-2} \xrightarrow{\vartheta_3} S_{k-1} \otimes G \xrightarrow{\vartheta_2} S_k \xrightarrow{\vartheta_1} R \to 0$$

of R/I^k we get a complex

$$D(k) : 0 \to R \to D_k \to D_{k-1} \otimes G^* \to D_{k-2} \to 0$$

with

$$H_0(D(k)) \simeq \text{Ext}^3(I^{(k)}/I^k, R) \simeq I^{(k)}/I^k$$

Since $C(k, \mathbf{x})$ is a resolution of $I^{(k)}/I^k$ there exists a comparison map

$$\alpha. : D(k) \to C(k, \mathbf{x}) \text{ with } \alpha_0 = id_{D_{k-2}}$$

In particular, we get a map

$$\alpha_3 : R \to S_{k-2}$$

Proposition 3.5 *Suppose I satisfies the conditions of 3.1, and let*

$$\alpha_3(1) = \sum_{\nu \in X} u_\nu e_1^{\nu_1} e_2^{\nu_2} e_3^{\nu_3}$$

with $X = \{\nu = (\nu_1, \nu_2, \nu_3)| \sum_{i=1}^{3} \nu_i = k - 2\}$. Then $\{u_\nu + I^k | \nu \in X\}$ is a minimal set of generators of $I^{(k)}/I^k$.

PROOF. The inclusion $j : I^{(k)}/I^k \to R/I^k$ induces a complex homomorphism $\beta : C(k, \mathbf{x}) \to S(k)$, and thus a complex homomorphism $\beta^* : D(k) \to C(k, \mathbf{x})$. Write $\beta^*(1) = \sum_{\nu \in X} v_\nu e_1^{\nu_1} e_2^{\nu_2} e_3^{\nu_3}$. It is clear from the definition of β that $\{v_\nu + I^k | \nu \in X\}$ is a minimal set of generators of $I^{(k)}/I^k$.

As

$$H_0(\beta_3^*) = \operatorname{Ext}^3(j, R) = id_{I^{(k)}/I^k}$$

we see that α and β^* are homotopic. Hence there exists a homomorphism $\varphi : D_k \to S_{k-2}$ such that $\alpha - \beta^* = \varphi \circ \vartheta_1^*$. But $\operatorname{Im}\varphi \circ \vartheta_1^* \subseteq I^k S_{k-2}$, and therefore $u_\nu + I^k = v_\nu + I^k$ for all $\nu \in X$. $\square$

Example 3.6 a) We know from 3.4 that $I^{(2)}/I^2$ has one generator $u + I^2$. We want to compute this generator in terms of the relation matrix

$$A = \begin{pmatrix} x_1 & x_2 & x_3 \\ y_1 & y_2 & y_3 \end{pmatrix}$$

under the assumption that $I_1(A) = (x_1, x_2, x_3)$.

According to 3.5 we have to compute α_3 in the commutative diagram

$$
\begin{array}{ccccccccc}
0 & \longrightarrow & R & \xrightarrow{\vartheta_1^*} & D_2 & \xrightarrow{\vartheta_2^*} & D_1 \otimes G^* & \xrightarrow{\vartheta_3^*} & D_0 \\
 & & \downarrow \alpha_3 & & \downarrow \alpha_2 & & \downarrow \alpha_1 & & \| \\
0 & \longrightarrow & S_0 & \xrightarrow{\partial_3} & S_1 & \xrightarrow{\partial_2} & D_1 & \xrightarrow{\partial_2} & D_0
\end{array}
$$

Notice that I is generated by d_1, d_2 and d_3, where

$$d_1 = x_2 y_3 - x_3 y_2$$

$$d_2 = x_3 y_1 - x_1 y_3$$

$$d_3 = x_1 y_3 - x_3 y_1$$

Let

$$y_i = \sum_{j=1}^{3} a_{ij} x_j \text{ for } i = 1, 2, 3$$

then α_1 and α_2 can be chosen as follows:

$$\alpha_1(f_i \otimes g_1^*) = -\sum_{j=1}^{3} a_{ij} f_j$$
$$\alpha_1(f_i \otimes g_2^*) = f_i$$

and

$$\alpha_2(f_1^{(2)}) = -a_{12} e_3 + a_{13} e_2$$
$$\alpha_2(f_2^{(2)}) = a_{21} e_3 - a_{23} e_1$$
$$\alpha_2(f_3^{(2)}) = -a_{31} e_2 + a_{32} e_1$$

$$\alpha_2(f_1 f_2) = -a_{22} e_3 + a_{23} e_2 + a_{11} e_3 - a_{13} e_1$$
$$\alpha_2(f_1 f_3) = -a_{32} e_3 + a_{33} e_2 - a_{11} e_2 + a_{12} e_1$$
$$\alpha_2(f_2 f_3) = a_{31} e_3 - a_{33} e_1 - a_{21} e_2 + a_{22} e_1$$

Now we find the generator $u + I^2$ of $I^{(2)}/I^2$ by comparing the coefficients in the following equation:

$$u x_1 e_1 + u x_2 e_2 + u x_3 e_3 = (\partial_3 \circ \alpha_3)(1) =$$
$$(\alpha_2 \circ \vartheta_1^*)(1) = \alpha_2(d_1^2 f_1^{(2)} + d_2^2 f_2^{(2)} + d_3^2 f_3^{(2)} + d_1 d_2 f_1 f_2 + d_1 d_3 f_1 f_3 + d_2 d_3 f_2 f_3) =$$
$$\Delta_1 e_1 + \Delta_2 e_2 + \Delta_3 e_3,$$

where the Δ_i are the maximal minors (with sign) of the matrix:

$$\begin{pmatrix} d_1 & d_2 & d_3 \\ \sum_i a_{i1} d_i & \sum_i a_{i2} d_i & \sum_i a_{i3} d_i \end{pmatrix}$$

Therefore we obtain

$$u = \Delta_i / x_i \text{ for } i = 1, 2, 3.$$

This is exactly the formula of Vasconcelos [9].

   b) We compute the three generators of $I^{(3)}/I^3$, if the relation matrix A of I is of 'monomial type' and satisfies the conditions of 3.1 up to $n = 3$. So

$$A = \begin{pmatrix} x_1 & x_2 & x_3 \\ x_3^2 a_1 & x_1^2 a_2 & x_2^2 a_3 \end{pmatrix}$$

We keep the notations of the previous example. If we compute the comparison map $\alpha_. : D(3) \to C(3, \mathbf{x})$ as before, we obtain $u_1, u_2, u_3 \in I^{(3)}$ such that
   1) $u_1 + I^3, u_2 + I^3, u_3 + I^3$ generate $I^{(3)}/I^3$.
   2) The u_i satisfy the following equations:

$$x_1 u_1 = \tfrac{1}{2}(a_3 d_3^3 + a_1 d_2^2 d_1)$$
$$x_2 u_2 = \tfrac{1}{2}(a_1 d_1^3 + a_2 d_3^2 d_2)$$
$$x_3 u_3 = \tfrac{1}{2}(a_2 d_2^3 + a_3 d_1^2 d_3)$$

$$x_1 u_2 + x_2 u_1 = -a_1 d_1^2 d_2$$
$$x_1 u_3 + x_3 u_1 = -a_3 d_3^2 d_1$$
$$x_2 u_3 + x_3 u_2 = -a_2 d_2^2 d_3$$

If char$R = 2$, then the resolution preceding Proposition 2.4 shows that $I^{(3)}/I^3$ is generated by one element, say $u + I^3$. Using this resolution we get:

$$x_1 x_2 x_3 u = x_1 a_2 d_2^2 d_3 + x_2 a_3 d_3^2 d_1 + x_3 a_1 d_1^2 d_2$$

and

$$\begin{array}{rcl}
x_1^2 u & = & a_3 d_3^3 + a_1 d_2^2 d_1 \\
x_2^2 u & = & a_1 d_1^3 + a_2 d_3^2 d_2 \\
x_3^2 u & = & a_2 d_2^3 + a_3 d_1^2 d_3
\end{array}$$

We conclude this paper with the following result which is an immediate consequence of 3.1 and of the arguments of 3.5. We leave its proof to the reader.

Corollary 3.7 *Suppose I satisfies the assumptions of 3.1. Then the mapping cone α_*^* : $C(k, \mathbf{x}) \to D(k)^*$ is an R-free resolution of $R/I^{(k)}$ for $k = 2, \ldots, n$.*

After having cancelled components in the mapping cone which split, one obtains the following R-free resolution

$$0 \to S_{k-1} \oplus D_{k-1} \xrightarrow{\varphi_k} S_k \oplus D_{k-2} \to R \to R/I^{(k)} \to 0$$

where

$$\varphi_k = \begin{pmatrix} \tilde{\vartheta}_2 & 0 \\ \alpha_2^* & \partial_1 \end{pmatrix}$$

and where $\tilde{\vartheta}_2$ is defined as the composition of $S_{k-1} \to S_{k-1} \otimes G$, $a \mapsto a \otimes g_1$ with $\vartheta_2 : S_{k-1} \otimes G \to S_k$.

The resolution is minimal if $\mathrm{Im}\alpha_2 \subseteq mS_{k-1}$.

References

[1] J.Herzog, A.Simis, W.V.Vasconcelos, On the arithmetic and homology of algebras of linear type, Trans. Amer. Math. Soc.**283** (1984), 661-683.

[2] J.Herzog, B.Ulrich, Self-linked curve singularities, Preprint (1988).

[3] C.Huneke, On the finite generation of symbolic blow-ups, Math. Z. **179** (1982), 465-472.

[4] C.Huneke, The primary components of and integral closures of ideals in 3-dimensional regular local rings, Math. Ann. **275** (1986), 617-635.

[5] M.Morales, Noetherian symbolic blow-up and examples in any dimension, Preprint(1987).

[6] P.Schenzel, Finiteness and noetherian symbolic blow-up rings, Proc. Amer. Math. Soc., to appear.

[7] P.Schenzel, Examples of noetherian symbolic blow-up rings, Rev. Roumaine Math. Pures Appl. **33** (1988) 4, 375-383.

[8] P.Roberts, A prime ideal in a polynomial ring whose symbolic blow-up is not Noetherian, Proc. Amer. Math. Soc. **94** (1985), 589-592.

[9] W.V.Vasconcelos, The structure of certain ideal transforms, Math. Z. **198** (1988), 435-448.

[10] W.V.Vasconcelos, Symmetric Algebras, Proceedings of the Microprogram in Commutative Algebra held at MSRI, Berkeley (1987), to appear.

[11] J.Weyman, Resolutions of the exterior and symmetric power of a module, J.Alg. **58** (1979), 333-341.

Universität Essen GHS, FB 6 Mathematik, Universitätsstraße 3, D-4300 Essen

Generic Residual Intersections

CRAIG HUNEKE* AND BERND ULRICH*

1. Introduction

In this paper we are concerned with residual intersections, a notion that essentially goes back to Artin and Nagata ([1]). Let X and Y be two irreducible closed subschemes of a Noetherian scheme Z with $\operatorname{codim}_Z X \leq \operatorname{codim}_Z Y = s$ and $Y \not\subset X$, then Y is called a residual intersection of X if the number of equations needed to define $X \cup Y$ as a subscheme of Z is smallest possible, namely s. However, in order to include the case where X and Y are reducible with X possibly containing some component of Y, we use the following more general definition:

DEFINITION 1.1 ([17], 1.1): Let R be a Noetherian ring, let I be an R-ideal, let $s \geq \operatorname{ht} I$, let $A = (a_1, \ldots, a_s) \subset I$ with $A \neq I$, and set $J = A : I$. If $\operatorname{ht} J \geq s$, then J is called an s–residual intersection of I. If furthermore $I_p = A_p$ for all $p \in V(I) = \{p \in \operatorname{Spec}(R) | I \subset p\}$ with $\dim R_p \leq s$, then J is called a geometric s–residual intersection of I.

Notice that residual intersection can be viewed as a natural generalization of linkage to the case where the two "linked" ideals need not have the same codimension. Indeed if R is a local Gorenstein ring and I is an unmixed R-ideal of grade g, then g–residual intersection corresponds to linkage and geometric g–residual intersection corresponds to geometric linkage.

Before we proceed, more definitions are needed. An ideal I in a Noetherian ring R is said to satisfy G_s if $\mu(I_p) \leq \dim R_p$ for all $p \in V(I)$ with $\dim R_p \leq s - 1$ (where μ denotes minimal number of generators), and G_∞ if I is G_s for every s ([1]). An ideal I in a local Cohen–Macaulay ring R is called strongly Cohen–Macaulay if all Koszul homology modules of some (and hence every) generating set of I are Cohen–Macaulay modules ([14]). Let (R, I) and (S, K) be pairs of Noetherian local rings R, S and ideals $I \subset R$, $K \subset S$, then (S, K) is said to be a deformation of (R, I), if $(R, I) = (S/(\underline{x}), (K, \underline{x})/(\underline{x}))$ for some sequence $\underline{x}$ in S which is regular on S and S/K. Properties of residual intersections and in particular their Cohen–Macaulayness have been studied in [1], [9], [14], [17]. We only quote the main result from [17]:

Theorem 1.2. ([17], 5.1) *Let R be a local Gorenstein ring, let I be an R–ideal of grade g, assume that (R, I) has a deformation (S, K) with K a strongly Cohen–Macaulay ideal satisfying G_∞, and let $J = A : I$ be an s–residual intersection of I.*

Then J is a Cohen–Macaulay ideal of grade s, $\operatorname{depth} R/A = \dim R - s$, and $\omega_{R/J}$, the canonical module of R/J, is isomorphic to $S_{s-g+1}(I/A)$, the $(s - g + 1)^{\text{th}}$ symmetric power of I/A.

The main focus in this paper is generic residual intersection, whose definition we now recall.

---

*Both authors were partially supported by the NSF.

DEFINITION 1.3 ([17], 3.1): Let R be a Noetherian ring, let $I = R$ with $s \geq 1$, or let $I \neq 0$ be an R–ideal satisfying G_{s+1} where $s \geq \max\{1, htI\}$. Further let $f_1, \ldots, f_n$ be a generating sequence of I, let X be a generic s by n matrix, and set

$$\begin{pmatrix} a_1 \\ \vdots \\ a_s \end{pmatrix} = X \begin{pmatrix} f_1 \\ \vdots \\ f_n \end{pmatrix}.$$

Then we define $RI(s; f_1, \ldots, f_n) = (a_1, \ldots, a_s)R[X] : IR[X]$ and call this $R[X]$–ideal the *generic s–residual intersection* of I with respect to $f_1, \ldots, f_n$. We will see that this definition is essentially independent of the chosen generating set of I (Lemma 2.2), and we will therefore often write $RI(s; I)$ instead of $RI(s; f_1, \ldots, f_n)$.

The main properties of generic residual intersections are proved in [17].

Theorem 1.4. ([17], 3.3) *Let R be a local Cohen–Macaulay ring, let I be a strongly Cohen–Macaulay R–ideal satisfying G_{s+1}, where $s \geq g = \mathrm{grade}\, I \geq 1$, in $S = R[X]$ consider a generic s–residual intersection $J = RI(s; I) = (a_1, \ldots, a_s)S : IS$ and write $A = (a_1, \ldots, a_s)S$, $\overline{I} = (IS + J)/J$. Then*

   a) *J is a geometric s–residual intersection of IS.*
   b) *(Follows from a) and [14], 3.1). J is a Cohen–Macaulay S–ideal of grade s, $A = IS \cap J$, and $\overline{I}$ is a strongly Cohen–Macaulay ideal of grade 1 in S/J.*
   c) *$\overline{I}$ satisfies G_∞ if I satisfies G_∞, $\omega_{S/J} \cong (\overline{I})^{s-g+1}$ if R is Gorenstein, and J is prime if R is a domain.*

The next result clarifies the relation between generic and arbitrary residual intersections, thus motivating the study of generic residual intersections.

Theorem 1.5. ([17], proof of 5.1). *Let R be a local Gorenstein ring, let I be a strongly Cohen–Macaulay ideal satisfying G_{s+1} where $s \geq \mathrm{grade}\, I \geq 1$, let $RI(s; I)$ be a generic s–residual intersection of I in $S = R[X]$, and let J be an arbitrary s–residual intersection of I in R.*
Then there exists $q \in \mathrm{Spec}(S)$ such that $(S_q, RI(s; I)_q)$ is a deformation of (R, J).

The above theorem means in particular that in the presence of rigidity assumptions on the residual intersection, generic residual intersection and arbitrary residual intersection coincide, at least up to localization and completion. To provide further motivation we are now going to list four rather natural classes of examples which all turn out to be (localizations of) generic residual intersections.

EXAMPLE 1.6 ([14]): Let R be a local Cohen–Macaulay ring, let $I = (f_1, \ldots, f_n)$ be a strongly Cohen–Macaulay ideal of grade $g \geq 1$ satisfying G_∞, let U be a variable, and consider a generic n–residual intersection $J = RI(n; f_1, \ldots, f_n, U)$ in $S = R[U, X, Y]$, with X a generic n by n matrix and Y a generic n by 1 matrix. Then $J = (a_1, \ldots, a_n)S : (I, U)S$, where

$$\begin{pmatrix} a_1 \\ \vdots \\ a_n \end{pmatrix} = (X|Y) \begin{pmatrix} f_1 \\ \vdots \\ f_n \\ U \end{pmatrix}.$$

Now over $R(X)$, the matrix X is invertible and we may perform elementary row and column operators to assume that

$$\begin{pmatrix} a_1' \\ \vdots \\ a_n' \end{pmatrix} = \left(\begin{array}{ccc|c} 1 & & 0 & T_1 \\ & \ddots & & \vdots \\ 0 & & 1 & T_n \end{array} \right) \begin{pmatrix} f_1' \\ \vdots \\ f_n' \\ U \end{pmatrix} = \begin{pmatrix} f_1' + T_1 U \\ \vdots \\ f_n' + T_n U \end{pmatrix}.$$

Here $R(X)[U, T_1, \ldots, T_n] = R(X)[U, Y]$, $T_1, \ldots, T_n$ are variables, $(f_1', \ldots, f_n')R(X) = IR(X)$ and $(a_1', \ldots, a_n')R(X)[U, T_1, \ldots, T_n] = (a_1, \ldots, a_n)R(X)[U, Y]$ (cf. the proof of Lemma 2.3 for more details). Therefore $(S \underset{R[X]}{\otimes} R(X), J \underset{R[X]}{\otimes} R(X)) = (R(X)[U, T_1, \ldots, T_n], (f_i' + T_i U | 1 \le i \le n) : (I, U))$.

On the other hand by [14], 4.3, $R(X)[U, T_1, \ldots, T_n]/((f_i' + T_i U | 1 \le i \le n) : (I, U)) \cong R(X)[It, t^{-1}]$, the extended Rees algebra of $IR(X)$. Thus it follows that

$$R[It, t^{-1}] \underset{R}{\otimes} R(X) \cong (S/RI(n; f_1, \ldots, f_n, U)) \underset{R[X]}{\otimes} R(X).$$

Moreover, $R[It, t^{-1}]/(t^{-1}) = R[It, t^{-1}]/(I, t^{-1}) \cong gr_I(R)$, the associated graded ring of I, and hence

$$gr_I(R) \underset{R}{\otimes} R(X) \cong (S/(U, RI(n; f_1, \ldots, f_n, U))) \underset{R[X]}{\otimes} R(X)$$
$$= (S/(I, U, RI(n; f_1, \ldots, f_n, U))) \underset{R[X]}{\otimes} R(X).$$

EXAMPLE 1.7 ([14]): Let R be a local Cohen–Macaulay ring, let Z be a generic $r - 1$ by r matrix ($r \ge 2$) with maximal minors $\Delta_1, \ldots, \Delta_r$, let Y be a generic $s - r + 1$ by r matrix ($s > r$), and let X be the generic s by r matrix ($\frac{Z}{Y}$). Now consider the generic $(s - r + 1)$–residual intersection $RI(s - r + 1; \Delta_1, \ldots, \Delta_r) \subset R[X]$ of the generic perfect grade 2 ideal $I_{r-1}(Z) \subset R[Z]$. Then

$$I_r(X) = RI(s - r + 1; \Delta_1, \ldots, \Delta_r)$$

([14], 4.1).

EXAMPLE 1.8 ([17]): Let R be a local Cohen–Macaulay ring, let $y_1, \ldots, y_g$ be a regular sequence in R, let X be a generic s by g matrix ($s \ge g$), and let $a_1, \ldots, a_s$ be the entries of the product matrix

$$X \begin{pmatrix} y_1 \\ \vdots \\ y_g \end{pmatrix}.$$

Then in $R[X]$,

$$(a_1, \ldots, a_s, I_g(X)) = RI(s; y_1, \ldots, y_g)$$

([17], 3.4). The resolution of such ideals was worked out in [4].

EXAMPLE 1.9 ([21]): Let R be a local Cohen–Macaulay ring, let X be a generic alternating $2n + 1$ by $2n + 1$ matrix ($n \ge 1$), let Y be a generic s by $2n + 1$ matrix ($s \ge 3$), and in $R[X, Y]$, consider the ideal P generated by all Pfaffians of all possible sizes containing X of the alternating matrix

$$\left(\begin{array}{c|c} X & -Y^t \\ \hline Y & 0 \end{array} \right).$$

Further let $p_1, \ldots, p_{2n+1}$ be the $2n$ by $2n$ Pfaffians of X (which generate a generic perfect Gorenstein ideal of grade 3 in $R[X]$ in case R is Gorenstein). Then

$$P = RI(s; p_1, \ldots, p_{2n+1})$$

([21]). The resolution of such ideals was worked out in [21].

Theorem 1.5 and Examples 1.6 through 1.9 provide applications for the results proved in the next two sections of this paper. There, generalizing known facts about generic linkage (in particular [15], 2.9 and 2.15), we study the singular locus and the divisor class group of generic residual intersections. More specifically, we give lower bounds for the codimension of the singular locus and the non–complete–intersection locus of algebras defined by generic residual intersections (Theorem 2.4), and we identify a generator of the divisor class group, which turns out to be an infinite cyclic group, (Theorem 3.4). Furthermore we classify all rank one Cohen–Macaulay modules over generic residual intersections (Theorem 3.5) (for generic linkage, this was done in [23]).

2. The Singular Locus

Before proving the main result in this section (Theorem 2.4), we first need several more definitions and lemmas.

DEFINITION 2.1 ([11]): Let (R, I) and (S, K) be pairs where R, S are Noetherian rings, and $I \subset R, K \subset S$ are ideals or $I = R$ or $K = S$. We say that (R, I) and (S, K) are *equivalent* and write $(R, I) \equiv (S, K)$ if there are finite sets of variables X over R and Z over S, and an isomorphism $\varphi : R[X] \to S[Z]$ such that $\varphi(IR[X]) = KS[Z]$.

Lemma 2.2. *Let R, I, s be as in Definition 1.3, choose two generating sequences $f_1, \ldots, f_n$ and $h_1, \ldots, h_m$ of I, let X be a generic s by n matrix, and let Z be a generic s by m matrix.*

Then $(R[X], RI(s; f_1, \ldots, f_n)) \equiv (R[Z], RI(s; h_1, \ldots, h_m))$. Moreover the isomorphism defining this equivalence is R-linear.

PROOF: Taking the union of the two generating sets and using induction, we may assume that $h_1, \ldots, h_m = f_1, \ldots, f_n, h$. In particular, $h = \sum_{j=1}^{n} r_j f_j$ for some $r_j \in R$. Now define an isomorphism of R-algebras $\varphi : R[X, Z] \to R[Z, X]$ by setting $\varphi(x_{ij}) = z_{ij} + r_j z_{in+1}$ for $1 \leq i \leq s$, $1 \leq j \leq n$, $\varphi(z_{ij}) = x_{ij}$ for $1 \leq i \leq s$, $1 \leq j \leq n$, and $\varphi(z_{in+1}) = z_{in+1}$ for $1 \leq i \leq s$. Then $\varphi(\sum_{j=1}^{n} x_{ij} f_j) = \sum_{j=1}^{n} (z_{ij} + r_j z_{in+1}) f_j = \sum_{j=1}^{n} z_{ij} f_j + z_{in+1} \sum_{j=1}^{n} r_j f_j = \sum_{j=1}^{n} z_{ij} f_j + z_{in+1} h$. Thus if we write

$$\left(\begin{array}{c} a_1 \\ \vdots \\ a_s \end{array} \right) = X \left(\begin{array}{c} f_1 \\ \vdots \\ f_n \end{array} \right), \quad \left(\begin{array}{c} b_1 \\ \vdots \\ b_s \end{array} \right) = Z \left(\begin{array}{c} f_1 \\ \vdots \\ f_n \\ h \end{array} \right)$$

then $\varphi(a_i) = b_i$ for all $1 \leq i \leq s$. Since moreover $\varphi(I) = I$, it follows that

$$\varphi(RI(s; f_1, \ldots, f_n)R[X, Z]) = \varphi((a_1, \ldots, a_s)R[X, Z] : IR[X, Z])$$
$$= \varphi((a_1, \ldots, a_s)R[X, Z]) : \varphi(IR[X, Z])$$
$$= (b_1, \ldots, b_s)R[Z, X] : IR[Z, X]$$
$$= RI(s; f_1, \ldots, f_n, h)R[Z, X]. \quad \blacksquare$$

In the light of Lemma 2.2 we will often write $RI(s; I)$ instead of $RI(s; f_1, \ldots, f_n)$. It is also clear that if W is a multiplicatively closed subset of R with $I_W \neq 0$, then

$$(R_W[X], RI(s; I_W)) \equiv (R[X]_W, RI(s; I)_W).$$

Lemma 2.3. *Let R be a local Cohen–Macaulay ring, let $I = (f_1, \ldots, f_g)$ be a complete intersection R-ideal of grade g, $s \geq g \geq 1$, consider the generic s-residual intersection $J = RI(s; f_1, \ldots f_g)$ in $S = R[X]$ where X is a generic s by g matrix, and let $q \in V(IS + J)$ with $\dim S_q \leq 2s - g + 3$.*

    a) *Then J_q and $(IS + J)_q$ are complete intersections.*
    b) *If in addition R/I is regular, then $(S/J)_q$ and $(S/IS + J)_q$ are regular.*

PROOF: Suppose that $I_{g-1}(X) \subset q$, then $IS + I_{g-1}(X) \subset q$, and hence $\dim S_q \geq g + 2(s - (g - 1) + 1) = 2s - g + 4$, which is ruled out by our assumption. Thus we may assume that $\Delta = \det U \notin q$ where U is a $g - 1$ by $g - 1$ matrix with

$$X = \left(\begin{array}{c|c} U & V \\ \hline W & Z \end{array} \right).$$

Hence there are invertible matrices A and B over the ring $R[U, \Delta^{-1}, V, W]$ such that

$$AXB = X' = \left(\begin{array}{ccc|c} 1 & & & 0 \\ & \ddots & & \vdots \\ & & 1 & 0 \\ \hline & & & Y_g \\ & 0 & & \vdots \\ & & & Y_s \end{array} \right),$$

and $R[U, \Delta^{-1}, V, W][Y_g, \ldots, Y_s] = S[\Delta^{-1}]$. In particular, $Y_g, \ldots, Y_s$ are variables over $R' = R[U, \Delta^{-1}, V, W]_{q^c}$, and $S_q = R'[Y_g, \ldots, Y_s]_q$.

Define

$$\left(\begin{array}{c} f_1' \\ \vdots \\ f_g' \end{array} \right) = B^{-1} \left(\begin{array}{c} f_1 \\ \vdots \\ f_g \end{array} \right),$$

$$\left(\begin{array}{c} a_1 \\ \vdots \\ a_s \end{array} \right) = X \left(\begin{array}{c} f_1 \\ \vdots \\ f_g \end{array} \right),$$

and

$$\begin{pmatrix} a'_1 \\ \vdots \\ a'_s \end{pmatrix} = X' \begin{pmatrix} f'_1 \\ \vdots \\ f'_g \end{pmatrix} = \begin{pmatrix} f'_1 \\ \vdots \\ f'_{g-1}, \\ Y_g f'_g \\ \vdots \\ Y_s f'_g \end{pmatrix},$$

Notice that

$$\begin{pmatrix} a'_1 \\ \vdots \\ a'_s \end{pmatrix} = A \begin{pmatrix} a_1 \\ \vdots \\ a_s \end{pmatrix}.$$

Now by Example 1.8, $J = (a_1, \ldots, a_s, I_g(X))S$. Therefore since $R' \subset S_q$ and A, B are invertible over R',

$$JS_q = (a'_1, \ldots, a'_s, I_g(X'))S_q$$
$$= (f'_1, \ldots, f'_{g-1}, Y_g, \ldots, Y_s)S_q.$$

Now our assertions follow since $S_q = R'[Y_g, \ldots, Y_s]_q$, $IR' = (f'_1, \ldots, f'_g)R'$, and $Y_g, \ldots, Y_s$ are variables over R'. ∎

Before stating our theorem, we need to agree on some more notation. Let R be a Noetherian ring and let I be an R-ideal. We say that I (or R/I, in case R is regular) satisfies (CI_k) if I_p is a complete intersection for all $p \in V(I)$ with $\dim(R/I)_p \leq k$. Furthermore one says that R satisfies Serre's condition (R_k) if R_p is regular for all $p \in \operatorname{Spec}(R)$ with $\dim R_p \leq k$.

Theorem 2.4. *Let R be a local Cohen–Macaulay ring, let I be a strongly Cohen–Macaulay R-ideal, $s \geq g = \operatorname{grade} I \geq 1$, $k \geq 0$, assume that $\mu(I_p) \leq \max\{g, \dim R_p - k\}$ for all $p \in V(I)$ with $\dim R_p \leq s + k$, consider a generic s-residual intersection $J = RI(s; I)$ in $S = R[X]$, and define $\ell = \min\{k, s - g + 3\}$.*

a) *Then J satisfies (CI_ℓ), and $IS + J$ satisfies $(CI_{\ell-1})$.*
b) *If R/I satisfies (R_{k-1}) and R_p is regular for all $p \in \operatorname{Spec}(R) \backslash V(I)$ with $\dim R_p \leq s + k$, then S/J satisfies (R_ℓ). If R/I satisfies (R_{k-1}), then $S/(IS + J)$ satisfies $(R_{\ell-1})$.*
c) *If $k \geq 1$ and I is prime, then $IS + J$ is prime.*

PROOF: Notice that by Theorem 1.4.b, J and $IS + J$ are unmixed S-ideals of grade s and $s + 1$ respectively.

We first show the following claim: Let $q \in V(J)$ with $\dim S_q \leq s + k$ and $q \cap R \in V(I)$, then $\dim R_{q \cap R} \leq g + k - 1$ and $I_{q \cap R}$ is a complete intersection. To prove this assertion, we may localize with respect to the multiplicative set $R \backslash (q \cap R)$ (cf. the remark following Lemma 2.2), and then assume that $q \cap R = m$, the maximal ideal of R. We need to show that $\dim R \leq g + k - 1$. Let $f_1, \ldots, f_n$ be any minimal generating set of I, then by Lemma 2.2, $(S, J) \equiv (R[Y], RI(s; f_1, \ldots, f_n))$ with Y a generic s by n matrix. Furthermore, $(S, IS + J) \equiv (R[Y], IR[Y] + RI(s; f_1, \ldots, f_n))$. Let q' be the prime ideal corresponding to q under the above equivalence, then $\dim R[Y]_{q'} \leq \dim S_q$, and $q' \cap R = m$ since the equivalence leaves R fixed. Thus changing back to our original notation, we may assume that $J = RI(s; f_1, \ldots, f_n)$ and $S = R[X]$ with X a generic s by n matrix. Write

$$\begin{pmatrix} a_1 \\ \vdots \\ a_s \end{pmatrix} = X \begin{pmatrix} f_1 \\ \vdots \\ f_n \end{pmatrix},$$

then $J = \operatorname{ann}_S((f_1, \ldots, f_n)S/(a_1, \ldots, a_s)S) \supset I_n(X)$. Thus $q \supset I_n(X) + m$, so that $s + k \geq \dim S_q \geq \dim R + s - n + 1$. Therefore $n \geq \dim R - k + 1$. On the other hand, $\dim R \leq \dim S_q \leq s + k$, and hence by our hypothesis $n = \mu(I) \leq \max\{g, \dim R - k\}$. Putting these two inequalities together yields that $n = g$ and $\dim R \leq g + k - 1$.

We are now ready to prove parts a) and b). Since J and $IS + J$ are unmixed of grade s and $s + 1$ respectively and $\ell \leq k$, it suffices to consider prime ideals $q \in V(J)$ and $q \in V(IS + J)$ respectively with $\dim S_q \leq s + k$. As before, we may assume that $q \cap R = m$ and that $J = RI(s; f_1, \ldots, f_n)$ where $f_1, \ldots, f_n$ is any minimal generating set of I. If $I = R$, then $J = RI(s; 1) = (Y_1, \ldots, Y_s)S$ with $Y_1, \ldots, Y_s$ variables over R, and our assertions follow easily. If however $I \neq R$, then the statement from the beginning of this proof implies that I is complete intersection and $\dim R \leq g + k - 1$. In particular $k \geq 1$, and R/I is regular (for part b). Now the claims in a) and b) follow from Lemma 2.3.

To prove part c), choose $k = 1$ and let $q \in \operatorname{Ass}(S/IS + J)$. Then $\dim S_q \leq s + 1 = s + k$, and $q \cap R \in V(I)$. Hence by the result from the beginning of this proof, $\dim R_{q \cap R} \leq g + k - 1 = g$, and therefore $q \cap R = I$. Thus

$$\left(\bigcup_{q \in \operatorname{Ass}(S/IS+J)} q \right) \cap R = I,$$

and it suffices to show that $(IS + J)_W$ is prime with $W = R \setminus I$. Hence localizing with respect to W and using Lemma 2.2, we may assume that $J = RI(s; f_1, \ldots, f_g)$ where $f_1, \ldots, f_g$ is a regular system of parameters of R, and $S = R[X]$ with X a generic s by g matrix. But then by Example 1.8, $IS + J = (f_1, \ldots, f_g, I_g(X))S$ which is obviously prime. ∎

As an illustration of how Theorem 2.4 can be applied to Examples 1.6 through 1.9, we now give a new proof of a well known result from [12] and [22].

Corollary 2.5. *Let R be a local Cohen–Macaulay ring, let I be a strongly Cohen–Macaulay ideal of grade g, let $k \geq 0$, and assume that $\mu(I_p) \leq \max\{g, \dim R_p - k\}$ for all $p \in V(I)$.*

    a) *If R is regular, then $R[It, t^{-1}]$ satisfies (CI_k), and $\operatorname{gr}_I(R)$ satisfies (CI_{k-1}).*

    b) *(Follows from combining [12], Proposition 2.1, and [8], 2.6). If R is regular and R/I satisfies (R_{k-1}), then $R[It, t^{-1}]$ satisfies (R_k), and $\operatorname{gr}_I(R)$ satisfies (R_{k-1}).*

    c) *([22], 3.4, or [12], Corollary 2.1, and [8], 2.6). If $k \geq 1$ and I is prime, then $\operatorname{gr}_I(R)$ is a domain.*

PROOF: We may assume that $g \geq 1$. Now the assertions follow from Example 1.6 and Theorem 2.4 (notice that we may choose $s = n$ to be arbitrarily large in Example 1.6, hence $\ell = k$ in Theorem 2.4). ∎

Let I be an ideal in a regular local ring R, then we say that (R, I) is *smoothable in codimension* k if (R, I) has a deformation (S, K) with char $S =$ char R and S/K satisfying (R_k).

Corollary 2.6. *Let R be a regular local ring, let $I \neq 0$ be an R-ideal, $s \geq g =$ grade I, $k \geq 0$, assume (R, I) has a deformation (S, K) with char $S =$ char R such that K is strongly Cohen-Macaulay, and $\mu(K_p) \leq \max\{g, \dim S_p - k\}$ for all $p \in V(K)$ with $\dim S_p \leq s + k$. Furthermore let J be any (not necessarily geometric or generic) s-residual intersection of I in R.*

Then (R, J) is smoothable in codimension $\min\{k, s - g + 3\}$.

PROOF: By [16], 3.10, we may deform (S, K) further to assume that S/K satisfies (R_k), and then by Theorem 2.4.b, $S[X]/RI(s; K)$ satisfies (R_ℓ) with $\ell = \min\{k, s - g + 3\}$. On the other hand, by [17], the proof of 5.1, there exists $q \in \operatorname{Spec}(S[X])$ such that $(S[X]_q, RI(s; K)_q)$ is a deformation of (R, J) (this is a stronger version of Theorem 1.5). ∎

REMARK 2.7: Let R be a regular local ring and let I be a licci R-ideal (i.e., an ideal admitting a sequence of links, $I = I_0 \sim I_1 \sim \cdots \sim I_n$ with I_n a complete intersection). Then I satisfies the assumptions of Corollary 2.6 with $k = 3$ ([13], 1.14, and [17], proof of 5.3), and $k = 5$ in case I is Gorenstein and the residue class field of R is infinite ([13], 1.14, [18]).

3. The Divisor Class Group

We first need a lemma that improves on Theorem 1.4.c.

Lemma 3.1. *Let R be a local Cohen-Macaulay ring, let I be an R-ideal, $s \geq g =$ grade $I \geq 1$, $0 \leq k \leq s - g + 2$, assume that $\mu(I_p) \leq \max\{g, \dim R_p - k\}$ for all $p \in V(I)$, consider a generic s-residual intersection $J = RI(s; I)$ in $S = R[X]$, and write $\overline{I} = (IS + J)/J$.*

Then $\mu(\overline{I}_q) \leq \max\{1, \dim S_q - s - k\}$ for all $q \in V(IS + J)$.

PROOF: Let $q \in V(IS + J)$ and localize at $q \cap R$ to assume that $q \cap R = m$, the maximal ideal of R. As in the proof of Lemma 2.3, we further reduce to the case where $J = RI(s; f_1, \ldots, f_n)$ with $f_1, \ldots, f_n$ a minimal generating set of I, and $S = R[X]$ with X a generic s by n matrix. Then $J = (a_1, \ldots, a_s)S : (f_1, \ldots, f_n)S$, where

$$\begin{pmatrix} a_1 \\ \vdots \\ a_s \end{pmatrix} = X \begin{pmatrix} f_1 \\ \vdots \\ f_n \end{pmatrix},$$

and certainly $\mu(\overline{I}_q) \leq \mu(((f_1, \ldots, f_n)S/(a_1, \ldots, a_s)S)_q)$.

Let $0 \leq t \leq \min\{n, s\}$ be minimal with $I_{t+1}(X) \subset q$. Then $I_t(X) \not\subset q$, and therefore

$$\mu(((f_1, \ldots, f_n)S/(a_1, \ldots, a_s)S)_q) \leq n - t.$$

On the other hand, since $(m, I_{t+1}(X)) \subset q$, we know that

$$\dim S_q \geq \dim R + (n - t)(s - t).$$

Thus it suffices to prove that $n - t \leq \max\{1, \dim R + (n - t)(s - t) - s - k\}$. Hence assuming $n - t \geq 2$, we need to show that

$$(3.2) \qquad\qquad n - t \leq \dim R + (n - t)(s - t) - s - k.$$

First consider the case $n \leq \dim R - k$. Then (3.2) follows once we have shown that

$$n - t \leq n + (n - t)(s - t) - s$$

or equivalently,

$$s - t \leq (n - t)(s - t),$$

which is obviously satisfied since $s - t \geq 0$ and $n - t \geq 2$.

Now consider the case where $n > \dim R - k$. Then by our assumption on I, we know that $n = g$. Since $\dim R \geq g$, (3.2) follows from the inequality

$$g - t \leq g + (g - t)(s - t) - s - k,$$

which is equivalent to

$$k \leq (g - t - 1)(s - t),$$

or

$$k \leq (g - t - 1)(s - g + 1 + g - t - 1).$$

However, the latter inequality holds since $g - t \geq 2$, hence $g - t - 1 \geq 1$, and $k \leq s - g + 2$ by assumption. ∎

Lemma 3.3. *Let R be a normal local Cohen–Macaulay domain, let $I = (f_1, \ldots, f_n)$ be a prime ideal of height one in R, let $Y_1, \ldots, Y_n$ be variables over R, and set $T = R[Y_1, \ldots, Y_n]$, $a = \sum_{i=1}^{n} Y_i f_i$.*

Then the divisor class group of $R, C\ell(R)$, is generated by the class of $I, [I]$, if and only if T_a is factorial.

PROOF: We only need to show that if T_a is factorial, then $C\ell(T) = \mathbf{Z}[IT]$.

Set $J = aT : IT$ (which is $RI(1; f_1, \ldots, f_n)$). Then by [15], 2.5 and 2.6, $IT \cap J = aT$, and J (as well as IT) is a prime ideal. But then Nagata's Lemma ([7], 7.1) yields an exact sequence

$$0 \rightarrow \mathbf{Z}[IT] + \mathbf{Z}[J] \rightarrow C\ell(T) \rightarrow C\ell(T_a) \rightarrow 0$$

where $C\ell(T_a) = 0$ and $[J] = -[IT]$. ∎

We are now ready to prove the second main theorem of this paper.

Theorem 3.4. *Let R be a local Cohen-Macaulay factorial domain, let I be a strongly Cohen-Macaulay prime ideal in R, $s \geq g = \operatorname{grade} I \geq 1$, and assume that $\mu(I_p) \leq \max\{g, \dim R_p - 1\}$ for all $p \in V(I)$ with $\dim R_p \leq s + 1$ and that R_p is regular for all $p \in \operatorname{Spec}(R) \backslash V(I)$ with $\dim R_p \leq s + 1$. Furthermore consider a generic s-residual intersection $J = RI(s; I)$ in $S = R[X]$, and write $\bar{I} = IS + J/J$.*

a) *Then $\bar{I}$ is a prime ideal of height one in S/J, S/J is normal, and $C\ell(S/J) = \mathbf{Z}[\bar{I}]$, where $[\bar{I}]$ has infinite order in case $g \geq 2$.*

b) *If $\mu(I_p) \leq \max\{g, \dim R_p - 1\}$ for all $p \in V(I)$, then $S_n(\bar{I}) \cong (\bar{I})^n = (\bar{I})^{(n)}$ for every $n \geq 1$ (where $(\bar{I})^{(n)}$ denotes the n-th symbolic power).*

PROOF: We first note that by Theorem 1.4.b, J is a Cohen-Macaulay ideal of grade s in S and $\bar{I}$ is a strongly Cohen–Macaulay ideal of height one in S/J (cf. also [14], 3.1). Furthermore, $\bar{I}$ is prime by Theorem 2.4.c.

We now prove part b). Since J is unmixed of grade s, since $ht\overline{I} = 1$, and since $\mu(I_p) \leq$ $\max\{g, \dim R_p - 1\}$, it follows from Lemma 3.1 that $\mu(\overline{I}_q) \leq \max\{ht\overline{I}, \dim(S/J)_q - 1\}$ for all $q \in V(IS + J)$. This together with the fact that $\overline{I}$ is a strongly Cohen–Macaulay prime ideal in a Cohen–Macaulay ring implies that $S_n(\overline{I}) \cong (\overline{I})^n$ ([8], 2.6), and that $(\overline{I})^n = (\overline{I})^{(n)}$ (Corollary 2.5.c, or [22], 3.4).

We are now ready to prove part a). We have already seen that $\overline{I}$ is a prime ideal of height one. Moreover, S/J is Cohen–Macaulay, and satisfies (R_1) by Theorem 2.4.b. In particular S/J is normal.

To show that $C\ell(S/J) = \mathbf{Z}[\overline{I}]$, we want to use Lemma 3.3. Let $f_1, \ldots, f_n$ be generators of I, let $Y_1, \ldots, Y_n$ be variables over S, and set $a = \sum_{i=1}^{n} Y_i f_i \in IR[Y_1, \ldots, Y_n]$. We need to prove that $(S/J)[Y_1, \ldots, Y_n]_a$ is factorial. However by Lemma 2.2,

$$(S[Y_1, \ldots, Y_n]_a, JS[Y_1, \ldots, Y_n]_a)$$
$$= (R[Y_1, \ldots, Y_n]_a[X], RI(s; IR[Y_1, \ldots, Y_n]_a))$$
$$\equiv (R[Y_1, \ldots, Y_n]_a[Z_1, \ldots, Z_s], RI(s; 1))$$

where $Z_1, \ldots, Z_s$ are variables and $RI(s; 1) = (Z_1, \ldots, Z_s)$. Hence

$$R[Y_1, \ldots, Y_n]_a[Z_1, \ldots, Z_s]/RI(s; 1) \cong R[Y_1, \ldots, Y_n]_a$$

is factorial, and therefore $(S/J)[Y_1, \ldots, Y_n]_a$ is also factorial.

It remains to prove that $g = 1$ in case $[\overline{I}]$ has finite order. So assume that $[\overline{I}]$ has finite order. Then the same holds true after localizing with respect to the multiplicative set $R\backslash I$, and we may assume that I is a complete intersection. In particular, part b) applies. But then $S_n(\overline{I}) \cong (\overline{I})^{(n)}$ is principal for some $n \geq 1$, which forces $\overline{I}$ and hence I to be principal. Therefore $g = 1$. ∎

Of course, the estimate on the local depth of $(\overline{I})^n$ in Theorem 3.4.b can be improved if we impose stronger conditions on the local number of generators of I (and use Lemma 3.1 and [8], 2.5). We are now able to classify the Cohen–Macaulay modules of rank one over $S/RI(s; I)$. For $s = \text{grade } I$, this was done in [23]. (We say that a finitely generated module M is Cohen–Macaulay if M localized at the irrelevant maximal ideal has this property.)

Theorem 3.5. *In addition to the assumptions and notations of Theorem 3.4.b, suppose that R is a factor ring of a local Gorenstein ring.*

Then M is a Cohen–Macaulay module of rank one over S/J if and only if M is isomorphic to $(\overline{I})^n$ for some $-1 \leq n \leq s - g + 2$ (where $(\overline{I})^{-1} \cong \text{Hom}(\overline{I}, S/J)$, $(\overline{I})^0 \cong S/J$).

PROOF: Since R is Gorenstein by [19], it follows from Theorem 1.4.c that $(\overline{I})^{s-g+1} \cong \omega$, the canonical module of S/J. Hence for every $n \geq 0$,

$$(\overline{I})^{-n} \cong \text{Hom}((\overline{I})^n, S/J) \cong \text{Hom}((\overline{I})^n \otimes \omega, \omega) \cong \text{Hom}((\overline{I})^n \otimes (\overline{I})^{s-g+1}, \omega)$$
$$\cong \text{Hom}((\overline{I})^{s-g+1+n}, \omega).$$

Therefore $\text{Hom}((\overline{I})^{-n}, \omega) \cong (\overline{I})^{s-g+1+n}$, because the latter module is reflexive by Theorem 3.4.b. Thus for $n \geq 0, (\overline{I})^{-n}$ is a Cohen–Macaulay module if the only if $(\overline{I})^{s-g+1+n}$

has this property. Also notice that $(\bar{I})^0$ and $\bar{I}$ are Cohen–Macaulay modules by Theorem 1.4.b. In the light of Theorem 3.4, Theorem 3.5 will follow once we have shown that for $g \geq 2$ and $n \in \mathbf{Z}$, $(\bar{I})^n$ is a Cohen–Macaulay module if and only if $-1 \leq n \leq s - g + 2$. By the above remarks it even suffices to prove the following three facts:

(3.6) $(\bar{I})^n$ is not a Cohen–Macaulay module for $n \geq s - g + 3$,
(3.7) $(\bar{I})^{-1}$ is a Cohen–Macaulay module,
(3.8) $(\bar{I})^n$ is a Cohen–Macaulay module for $2 \leq n \leq s - g + 1$.

To prove (3.6), suppose that $(\bar{I})^n$ is Cohen–Macaulay for some $n \geq s - g + 3$. Then the same holds true after localizing with respect to the multiplicative set $R \backslash I$. By Lemma 2.2 we further reduce to the situation where $J = RI(s; f_1, \ldots, f_g)$ in $S = R[X]$, with $f_1, \ldots, f_g$ a regular system of parameters of R and X a generic s by g matrix. Now localizing with respect to a $g - 2$ by $g - 2$ minor of X, changing the generators of I (cf. the proof of Lemma 2.3 for details), factoring out $g - 2$ of these generators, localizing R, deforming (which we are allowed to do by Theorem 1.4.a and [17], 4.2.ii), and localizing R again, we may assume that $R = k[Y_1, Y_2]_{(Y_1, Y_2)}$ with k a field, and Y_1, Y_2 variables, $J = RI(s - g + 2; Y_1, Y_2)$, and $S = R[X]$ with X a generic $s - g + 2$ by 2 matrix. Now write

$$\begin{pmatrix} a_1 \\ \vdots \\ a_{s-g+2} \end{pmatrix} = X \begin{pmatrix} Y_1 \\ Y_2 \end{pmatrix}.$$

Then by Theorem 1.4.b, $IS \cap J = (a_1, \ldots, a_{s-g+2})S$, hence $\bar{I} \cong (Y_1, Y_2)S/(a_1, \ldots, a_{s-g+2})S$. However, the latter module is the cokernel of the generic map $\varphi : S^{s-g+3} \rightarrow S^2$ given by the matrix

$$\begin{pmatrix} -Y_2 & Y_1 \\ X & \end{pmatrix},$$

and hence by Theorem 3.4.b, $(\bar{I})^n = S_n(\mathrm{coker}\,\varphi)$. On the other hand minimal free S-resolutions of the symmetric powers of $\mathrm{coker}\,\varphi$ were worked out in [6], [20], and it turns out that $S_n(\mathrm{coker}\,\varphi)$ is never Cohen–Macaulay for $n \geq (s - g + 3) - 2 + 2 = s - g + 3$. This finishes the proof of (3.6).

We are now going to show (3.7) and (3.8). To this end let $f_1, \ldots, f_n$ be a generating set of I, let X be a generic s by n matrix over R, and let

$$\begin{pmatrix} a_1 \\ \vdots \\ a_s \end{pmatrix} = X \begin{pmatrix} f_1 \\ \vdots \\ f_n \end{pmatrix},$$

so that $S = R[X]$ and $J = (a_1, \ldots, a_s)S : IS$. Furthermore let $Y_1, \ldots, Y_n$ be variables over S, set $T = S[Y_1, \ldots, Y_n]$, let "$\sim$" denote ideal generation in T/JT, let $a = \sum_{i=1}^{n} Y_i f_i$, and define $K = (a_1, \ldots, a_s, a)T : IT$. Then $K = RI(s + 1; f_1, \ldots, f_n)$ and $JT \subset K$. In particular, K is a Cohen–Macaulay ideal of grade $s + 1$ (Theorem 1.4.b). Hence $\widetilde{K}$ is a Cohen–Macaulay ideal of height one in a Cohen–Macaulay ring, and therefore $\widetilde{K}$ is a Cohen–Macaulay module. Moreover, $\widetilde{K} = (\tilde{a}) : \widetilde{I}$ by the proof of [14], 3.1, (and $\tilde{a}$ is regular on T/JT). Thus $(\bar{I})^{-1} \otimes_S T \cong \mathrm{Hom}(\widetilde{I}, T/JT) \cong \widetilde{K}$ is a Cohen–Macaulay module, and hence $(\bar{I})^{-1}$ is Cohen–Macaulay. This finishes the proof of (3.7).

We now prove (3.8) by induction on $s - g \geq 0$. For $s = g$, nothing is to be shown. So assume that the assertion holds for $s \geq g$ and $J = RI(s; I)$. To show the claim for $s + 1$, it suffices to consider the particular ideal $K = RI(s + 1; f_1, \ldots, f_n)$ from above, since any other $s+1$ residual intersection of I is equivalent to (T, K) and this equivalence leaves I^n fixed. We will denote ideal generation in T/K by "'". Because $(I')^{s-g+2}$ is the canonical module of T/K (Theorem 1.4.c), it actually suffices to prove that $(I')^n$ is a Cohen–Macaulay module for $2 \leq n \leq s - g + 1$. However this will follow once we have come up with an exact sequence

$$(3.9) \qquad\qquad 0 \to (\overline{I})^{n-1} \underset{S}{\otimes} T \to (\overline{I})^n \underset{S}{\otimes} T \to (I')^n \to 0$$

(notice that $\overline{I}$ is Cohen–Macaulay, $(\overline{I})^n$ is Cohen–Macaulay for $2 \leq n \leq s - g + 1$ by induction hypothesis, and $\dim I' = \dim \overline{I} - 1$ by Theorem 1.4.b). To obtain (3.9), we observe that $(\widetilde{a}) = \widetilde{I} \cap \widetilde{K}$ (since $\widetilde{K} = (\widetilde{a}) : \widetilde{I}$, cf. [14], proof of 3.1). Therefore, $(\widetilde{I})^n \cap \widetilde{K} = (\widetilde{I})^n \cap (\widetilde{I} \cap \widetilde{K}) = (\widetilde{I})^n \cap (\widetilde{a}) = \widetilde{a}(\widetilde{I})^{n-1}$, where the last equality follows from the fact that $\widetilde{a} \in \widetilde{I} \backslash (\widetilde{I})^2$ and $\widetilde{a}$ is regular on $gr_{\widetilde{I}}(T/JT)$ ($\widetilde{I}$ is SCM and G_∞, and grade $\widetilde{I} \geq 1$, hence $\widetilde{I}/(\widetilde{I})^2$ generates an ideal of positive grade in $gr_{\widetilde{I}}(T/JT)$, cf. [8], 2.6; on the other hand $\widetilde{a}$ is a general linear combination of elements in $\overline{I}/(\overline{I})^2$, cf. [11]). But then $(I')^n \cong (\widetilde{I})^n/(\widetilde{I})^n \cap \widetilde{K} = (\widetilde{I})^n/\widetilde{a}(\widetilde{I})^{n-1}$, where $(\widetilde{I})^n \cong (\overline{I})^n \underset{S}{\otimes} T$, $(\widetilde{I})^{n-1} \cong (\overline{I})^{n-1} \underset{S}{\otimes} T$, and $\widetilde{a}$ is regular on the latter module. This yields (3.9) and concludes the proof of (3.8). ∎

Corollary 3.10. *Let R be a regular local ring, let $I \neq 0$ be an R-ideal, $s \geq g = $ grade I, and assume that (R, I) has a deformation (S, K) where K is strongly Cohen–Macaulay and satisfies G_∞. Furthermore let $J = (a_1, \ldots, a_s) : I$ be a (not necessarily geometric or generic) s-residual intersection of I, and write $M = I/(a_1, \ldots, a_s)$, $\overline{I} = I + J/J$.*

Then $S_n(M)$ are maximal Cohen–Macaulay modules over S/J for $1 \leq n \leq s - g + 2$. If the residual intersection is geometric, then $S_n(M) \cong (\overline{I})^n$ for $1 \leq n \leq s - g + 2$.

PROOF: The first claim follows from a deformation argument as in the proof of Corollary 2.6, and from the proof of (3.8) (notice that (3.8) was proved only using the G_∞ property of I instead of the stronger assumption in Theorem 3.4.b). To see the second claim notice that for geometric residual intersections, $M \cong \overline{I}$ ([17], 5.1). ∎

When applied to Examples 1.6 through 1.9, Theorems 3.4 and 3.5 yield several new and many well known results concerning the divisor class group of various algebras ([2], [3], [10], [24], [25]) and the depth of symmetric powers of certain modules ([5], [6], [20]). We single out one of them.

EXAMPLE 3.11 ([2], [5], [6], [20], [24]): We use the notation from Example 1.7. In addition let R be a factorial Gorenstein domain and write $S = R[X]$, let

$$\begin{pmatrix} a_1 \\ \vdots \\ a_{s-r+1} \end{pmatrix} = Y \begin{pmatrix} \Delta_1 \\ \vdots \\ \Delta_r \end{pmatrix},$$

and let M be the cokernel of the generic map $S^s \to S^r$ given by the matrix X. Then $M \cong I_{r-1}(Z)S/(a_1, \ldots, a_{s-r+1})S$, since $I_{r-1}(Z)$ is the cokernel of Z and $X = (\frac{Z}{Y})$. Therefore $M \cong (I_{r-1}(Z)S + I_r(X))/I_r(X)$ (Theorem 1.4.b), $S_n(M)$ is reflexive for every

$n \geq 1$ (Theorem 3.4.b) and Cohen–Macaulay if and only if $1 \leq n \leq s - r + 1$ (Theorem 3.5). Furthermore the class of $\bar{I} = (I_{r-1}(Z)S + I_r(X))/I_r(X)$ generates the divisor class group of $S/I_r(X)$, which is an infinite cyclic group, (Theorem 3.4.a), $\bar{I}^{s-r}$ is the canonical module of $S/I_r(X)$ (Theorem 1.4.c), and $(\bar{I})^n$ is Cohen–Macaulay if and only if $-1 \leq n \leq s - r + 1$ (Theorem 3.5).

References

1. M. ARTIN AND M. NAGATA, *Residual intersections in Cohen–Macaulay rings*, J. Math. Kyoto Univ. **12** (1972), 307–323.

2. W. BRUNS, *Die Divisorenklassengruppe der Restklassenringe von Polynomringen nach Determinantenidealen*, Revue Roumaine Math. Pur. Appl. **20**(1975), 1109–1111.

3. W. BRUNS, *Divisors on varieties of complexes*, Math. Ann. **264** (1983), 53–71.

4. W. BRUNS, A. KUSTIN, AND M. MILLER, *The resolution of the generic residual intersection of a complete intersection*, preprint.

5. W. BRUNS AND U. VETTER, "Determinantal Rings," Lect. Notes Math. 1327, Springer, Berlin–Heidelberg, 1988.

6. D. BUCHSBAUM AND D. EISENBUD, *Remarks on ideals and resolutions*, Sympos. Math. **XI** (1973), 193–204.

7. R. FOSSUM, "The divisor class group of a Krull domain," Springer, Berlin–Heidelberg, 1973.

8. J. HERZOG, A. SIMIS, AND W. VASCONCELOS, *Approximation complexes and blowing-up rings*, J. Algebra **74** (1982), 466–493.

9. J. HERZOG, W. VASCONCELOS, AND R. VILLARREAL, *Ideals with sliding depth*, Nagoya Math. J. **99** (1985), 159–172.

10. M. HOCHSTER, *Criteria for the equality of ordinary and symbolic powers*, Math. Z. **133** (1973), 53–65.

11. M. HOCHSTER, *Properties of Noetherian rings stable under general grade reduction*, Arch. Math. **24** (1973), 393–396.

12. C. HUNEKE, *On the associated graded ring of an ideal*, Illinois J. Math. **26** (1982), 121–137.

13. C. HUNEKE, *Linkage and the Koszul homology of ideals*, Amer. J. Math. **104** (1982), 1043–1062.

14. C. HUNEKE, *Strongly Cohen–Macaulay schemes and residual intersections*, Trans. Amer. Math. Soc. **277** (1983), 739–673.

15. C. HUNEKE AND B. ULRICH, *Divisor class groups and deformations*, Amer. J. Math. **107** (1985), 1265–1303.

16. C. HUNEKE AND B. ULRICH, *Algebraic linkage*, Duke Math. J. **56** (1988), 415–429.

17. C. HUNEKE AND B. ULRICH, *Residual intersections*, J. reine angew. Math. **390** (1988), 1–20.

18. C. HUNEKE AND B. ULRICH, *Local properties of licci ideals*, in preparation.

19. P. MURTHY, *A note on factorial rings*, Arch. Math. **15** (1964), 418–420.

20. D. KIRBY, *A sequence of complexes associated with a matrix*, J. London Math. Soc. **7** (1973), 523–530.

21. A. KUSTIN AND B. ULRICH, *A family of complexes associated to an almost alternating map, with applications to residual intersections*, in preparation.

22. A. SIMIS AND W. VASCONCELOS, *The syzygies of the conormal module*, Amer. J. Math. **103** (1981), 203–224.

23. B. ULRICH, *Parafactoriality and small divisor class groups*, in preparation.

24. Y. YOSHINO, *The canonical module of graded rings defined by generic matrices*, Nagoya Math. J. **81** (1981), 105–112.

25. Y. YOSHINO, *Some results on the variety of complexes*, Nagoya J. Math **93** (1984), 39–60.

Department of Mathematics, Purdue University, West Lafayette, IN 47907, USA

Department of Mathematics, Michigan State University, East Lansing, MI 48824, USA

Subalgebra bases

LORENZO ROBBIANO AND MOSS SWEEDLER

Introduction

We present methods of computation about subalgebras of the polynomial ring. Previous work by Shannon and Sweedler [16] reduced subalgebra to problems in ideal theory and applied Buchberger's, now standard, Gröbner basis methods. The theory described here is an essentially different approach which does not reduce subalgebra questions to ideal questions. Theories of *ideal bases* and *subalgebra bases* may be approached from the ring theory/algebra view or the term rewriting/unification view. This article takes the ring theory/algebra view.

While subalgebras may not be as important as ideals, they are the second major type of *subobject* in ring theory. Gröbner bases answer the membership question and many other questions for ideals. A theory for subalgebras which is analogous to Buchberger's theory for ideals would directly answer the subalgebra membership question. To make matters *easier*, Buchberger's theory for ideals provides a prototype of what a theory of bases for subalgebras might look like. This program is carried out in the present paper. We present bases for subalgebras which are the **Subalgebra Analog to Gröbner Bases for Ideals (SAGBI)**. Unlike Gröbner bases, SAGBI bases are not always finite. This is not surprising because, unlike ideals in polynomial rings, subalgebras of polynomial rings are not necessarily finitely generated. Just as a Gröbner basis for an ideal generates the ideal, a SAGBI basis for a subalgebra generates the subalgebra.[1] Thus subalgebras which are not finitely generated can not have finite SAGBI bases. A more subtle issue determines whether a subalgebra has a finite SAGBI basis. It can happen that a finitely generated subalgebra has a finite SAGBI basis with respect to one term ordering, while with respect to another term ordering all SAGBI bases are infinite. It can also happen that a finitely generated subalgebra has no finite SAGBI basis, no matter the term ordering. Given a subalgebra B of a polynomial ring A, we compute an associated graded ring gr B with respect to a given term-ordering. The computation deals only with "algebra operations". The existence of a finite SAGBI basis for B comes down to whether gr B is finitely generated. We present an example where starting with a finitely generated subalgebra B, the finite generation of gr B depends on the choice of the term-ordering. However, in the case of subalgebras of $k[X]$ this phenomenon does not occur. Hence, for subalgebras of $k[X]$, our construction for finding a SAGBI basis for B, or equivalently for finding generators for gr B, is an algorithm. There are other aspects of the SAGBI theory which are algorithmic. For example, the subalgebra membership problem for homogeneous subalgebras is algorithmic.

The four cornerstones of Buchberger's constructive ideal theory are:

term orderings	test for a set to be a Gröbner basis
reduction	construction of a Gröbner basis from a set

Supported in part by the National Science Foundation and the U.S. Army Research Office through the Mathematical Sciences Institute of Cornell University.

[1] Of course, *generate* has a different meaning for ideals than for subalgebra.

SAGBI theory has similar cornerstones. Term orderings are *exactly the same* in both Buchberger theory and SAGBI theory. The other three cornerstones of SAGBI theory are analogous to the remaining cornerstones of Buchberger's theory. On the other hand, SAGBI theory is not simply a formal translation of Buchberger theory from ideals to subalgebras. For example, Buchberger's use of S-pairs does not simply translate to the subalgebra setting. The SAGBI replacement for S-pairs is a major deviation of SAGBI theory from Buchberger theory. Also, the SAGBI analog to one Gröbner basis construction method provides a SAGBI basis construction method while the SAGBI analog to another Gröbner basis construction method gives an incorrect SAGBI basis construction method.

The first great gift we can bestow on others is a good example[2]. In section 1 we present an example of a situation mentioned earlier. The example is a homogeneous subalgebra B of $k[X,Y]$ where $k[X,Y]$ is integral over B and B has no finite SAGBI basis, no matter the term ordering. Later we present an example of a homogeneous subalgebra B of $k[X,Y]$ where $k[X,Y]$ is integral over B and B has a finite SAGBI basis with one term ordering and no finite SAGBI basis with another term ordering.[3] We also present the example of the elementary symmetric functions being a SAGBI basis for the subalgebra of symmetric polynomials, with respect to any term ordering! These are but three of many examples presented herein.

Overview of the paper. The first section introduces basic notation and terminology including subduction[4] and the definition of SAGBI basis. There are many examples and the results include:

(1) SAGBI bases generate.
(2) Subduction over a SAGBI basis tests for subalgebra membership.
(3) Characterization of SAGBI basis of a subalgebra generated by one element.
(4) The elementary symmetric polynomials form a SAGBI basis with any term ordering.

The main features of section two are the notion of *tête-a-têtes* and the use of tête-a-têtes to determine whether a set is a SAGBI basis. *Tête-a-têtes* are the SAGBI replacement for Buchberger's S-pairs. Unlike S-pairs, in the SAGBI theory pair-wise relations are not adequate to test for a basis and tête-a-têtes are more than a formal analog to S-pairs. Modulo the difference between tête-a-têtes and S-pairs, the test for a basis in Buchberger theory and SAGBI theory are analogous. A main feature of section three is the SAGBI basis construction method. The method is not algorithmic in general. One way to recognize finite termination is presented but finer results about finite termination appear in section four. The SAGBI basis construction is analogous to a common Gröbner basis construction. Section three also presents the SAGBI analog to another common Gröbner basis construction method and shows it does not work! Section four presents the graded view of SAGBI theory. The grading associated with a term ordering plays the same critical role in SAGBI theory as in Buchberger theory. For example, a set S contained in a subalgebra B is a SAGBI basis for B if and only if the lead terms of S generate the

---

[2] Fortune cookie, Pan An Chinese Restaurant, Ithaca NY, 10/24/88.

[3] In this and the previously mentioned example, the integrality insures that B is finitely generated as an algebra.

[4] The SAGBI analog of Buchberger reduction.

associated graded algebra to B. As a corollary: a subalgebra has a finite SAGBI basis if and only if the associated graded algebra is finitely generated. Using the grading and results about monoids shows that our SAGBI basis construction method terminates in a finite number of steps if a subalgebra has a finite SAGBI basis. Section four also presents integrality conditions which insure a finite SAGBI basis and the corollary that SAGBI theory is algorithmic for subalgebras of $k[X]$. The use of the Hilbert function in SAGBI theory is illustrated in an example where finiteness of the SAGBI basis depends on the term ordering. Section four also includes a few words about graded structures. Section five is a brief section which shows that homogeneous SAGBI subalgebra membership determination is algorithmic. This rests on the fact that in SAGBI theory, just as in Buchberger theory, refinements are possible when dealing with homogeneous elements.

In the fall of 1988, after a talk about our work [14], Kapur told us of overlapping work by himself and Madlener from the term rewriting view. The Kapur, Madlener work was presented the following summer at the Computers and Mathematics conference in Cambridge MA and can be found in [6]. For a very general overview of the term rewriting/unification viewpoint see [2].

[9] presents related work giving a semi-decision procedure for computing in non-commutative polynomial rings.

1. Term orderings and SAGBI Reductions (Subduction)

1.1. NOTATION AND TERMINOLOGY: **N** denotes $\{0, 1, 2, \cdots\}$. If S and T are subsets of a common set, $S \setminus T$ denotes: $\{s \in S \mid s \notin T\}$. $k[K_1, \cdots, X_n]$ is frequently abbreviated: $k[\mathbf{X}]$. By a *term* in the polynomial ring $k[\mathbf{X}]$, we mean an element of the form: $X_1^{e_1} \cdots X_n^{e_n}$ for $e_1, \cdots, e_n \in \mathbf{N}$. A term times an element of k is called a *monomial*. An *exponent function on a commutative ring* R is a map from R to $\mathbf{N}$ which has finite support. If $E: R \to N$, *the support of* E *is*: $\{r \in R \mid E(r) \neq 0\}$. If E is an exponent function on R, R^E is defined as $\prod_{t \in T} t^{E(t)}$ where T is any finite subset of R containing the support of E.[5] If S is a subset of R containing the support of E we may write S^E for R^E and we may say that E is an exponent function on S. For $S \subset k[\mathbf{X}]$, an S *power product* is an element of $k[\mathbf{X}]$ of the form S^E where S contains the support of the exponent function E on $k[\mathbf{X}]$. PP S denotes the subset of $k[\mathbf{X}]$ consisting of S power products.

1.2. If $E, F: R \to \mathbf{N}$ are two exponent functions, the point-wise sum $E + F$ is an exponent function and $R^{E+F} = R^E R^F$. If $m \in \mathbf{N}$ then mE is the exponent function $E + \cdots + E$, m times and $R^{mE} = (R^E)^m$.

Here are the first two cornerstones of SAGBI theory.

1.3. TERM ORDERING: By *term ordering* we mean the usual notion of a well ordering of terms of a polynomial ring where all terms are greater than or equal to 1 and the product of terms preserves the order.[6] For more on term orderings see, [12] and [11].

1.4. LEAD TERM and LEAD MONOMIAL: When a term ordering has been specified on $k[\mathbf{X}]$ and $0 \neq f \in k[\mathbf{X}]$, LT f denotes the *lead term* of f with respect to the term

---

[5]In case E is the constant function *zero*, with the empty set as support, R^E is defined as 1.

[6]I.e.: $1 \leq \text{term}_1$ and $\text{term}_1 \leq \text{term}_2 \Rightarrow \text{term}_1 * \text{term}_3 \leq \text{term}_2 * \text{term}_3$ for all terms: term_1, term_2, term_3.

ordering. I.e. LT f is the term of f with non-zero coefficient which is greater than all other terms of f with non-zero coefficient. The *lead monomial* of f is this lead term of f times its coefficient. For a subset U of $k[\mathbf{X}]$, LT U denotes the set: $\{t \in k[\mathbf{X}] \mid t = \text{LT } f \text{ for } f \in U\}$.

Unless specified otherwise, $k[\mathbf{X}]$ is assumed to come equipped with a term ordering! LT has the following easy properties with respect to sums and products. Suppose $f_i \in k[\mathbf{X}]$ for $i = 1, \cdots, M$:

1.4.a If $\sum_1^M f_i$ is not zero and a non-zero f_h is chosen with maximal lead term among the non-zero f_i's: $\text{LT}(\sum_1^M f_i) \leq \text{LT } f_h$.

1.4.b If there is a unique non-zero f_h with maximal lead term among the non-zero f_i's: $\sum_1^M f_i$ is not zero and $\text{LT}(\sum_1^M f_i) = \text{LT } f_h$.

1.4.c If none of the f_i's are zero: $\text{LT}(\prod_1^M f_i) = \prod_1^M (\text{LT } f_i)$. This applies to S power products: for an S power product S^E, where E is an exponent function which does not contain zero in its support, $\text{LT } S^E = \prod_{s \in S} (\text{LT } s)^{E_s}$. This shows:

1.4.d: $$\text{LT}(\text{PP } S) = \text{PP}(\text{LT } S)$$

Suppose k is a field, we shall define the process of *SAGBI reduction* of an element of $k[\mathbf{X}]$ over a subset of $k[\mathbf{X}]$. In order to avoid calling the process *SAGBI reduction* to use the name *subduction*.[7]

1.5. SUBDUCTION: Suppose $b \in k[\mathbf{X}] \supset S$, and $k[\mathbf{X}]$ has a term ordering. *Subduction of b over S (with respect to the term ordering and k) is performed as follows:*

1.5.a: INITIALIZE: Set $b_0 = b$.

1.5.b: STOP ?: If $b_i \in k$, stop.

1.5.c: STOP ?: If LT b_i does not lie in $\text{PP}(\text{LT } S) = \text{LT}(\text{PP } S)$, stop.

1.5.d: Getting this far, i.e. not *stopping* at a previous step, implies that b_i does not lie in k, so LT $b_i \neq 1$, and LT b_i lies in $\text{PP}(\text{LT } S) = \text{LT}(\text{PP } S)$. Thus the next step makes sense.

1.5.e SUBDUCT: There is an exponent function E_i on S with LT $b_i = \text{LT } S^{E_i}$. Let $b_{i+1} = b_i - \gamma_i S^{E_i}$ where $\gamma_i \in k$ is chosen so that b_i and $\gamma_i S^{E_i}$ have the same lead monomial.

1.5.f: By choice of $\gamma_i : b_{i+1}$ equals zero or LT $b_{i+1} < \text{LT } b_i$.

1.5.g REPEAT: Go to step (1.5.b).

1.6. TERMINOLOGY: b_M is called an Mth *subductum of b over S* and we say that b *subduces* to b_M. We say that b *has a first subduction* or *first subductum over S* if (1.5) produces at least b_1. By (1.5.d): b has a first subduction over S if and only if b does not lie in k and LT b lies in $\text{PP}(\text{LT } S)$.

1.7 Remark. Notice that the exponent function (or S power product) at step (1.5.e) is not unique. This is comparable to the situation in Buchberger's reduction. For algorithmic purposes it is possible to specify a unique choice. For purposes of developing theory

---

[7] *Subalgebra reduction.*

it is better not to specify a choice at this point. The price of not specifying which power product to use is that b may have various subductions over S terminating at different elements. We shall be particularly interested whether subductions of b over S terminate at an element of k. We shall see that when S is a *SAGBI basis* (1.11), all subductions of b over S terminate at an element of k or no subductions of b over S terminate at an element of k (1.16.a).

1.8 Remark. At many points, e.g. (1.5.b), (1.11), we have statements which are conditional on elements *(not) lying in* k. Without changing the overall theory many of these conditions could be put in terms of elements *(not) being zero*. There are two reasons we use *(not) lying in* k rather than *(not) being zero*.

Subduction defined with respect to *(not) lying in* k usually stops one step sooner than subduction defined with respect to *(not) being zero*.

If there is a useful *relative* theory of SAGBI bases, relative to a general subalgebra C, (1.5.b) is likely to become: 1.5.b' STEP: If $b_i \in C$, stop.

By (1.10.e), (1.5) must terminate after a finite number of steps, even when S is infinite. Given a term T, it is a constructive matter to determine whether T is a power product of elements of a given finite set of terms. Thus when S is finite, and arithmetic can be performed constructively in k,[8] (1.5) is constructive.

1.9 Remark. An important part of subduction is the determination whether an element of $\mathbf{N}^n$ lies in a submonoid of $\mathbf{N}^n$ given by explicit generators. Moreover, if the answer is *yes*, the representation of the element in the submonoid is required, [1.5.c], [1.5.d], [1.5.e]. This is the problem of finding positive solutions to linear diophantine equations. Work on this and related problems can be found in: [3], [7], [10], [19], where additional references appear. Without concern for matters of efficiency, the determination of submonoid membership can be reduced to the determination of subalgebra membership. The latter is routinely handled by Gröbner basis methods. Suppose $t \in \mathbf{N}^n$, $S \subset \mathbf{N}^n$ and the question is to determine membership of t in the submonoid generated by S. Let T be the subset of S consisting of elements which are no larger than t in each component. T is a finite set and if t lies in the submonoid generated by S then t lies in the submonoid generated by T. Replacing S by T shows that S may be assumed to be finite. Let B be the subalgebra of $k[\mathbf{X}]$ generated by $\{\mathbf{X}^s\}_{s \in S}$. If t is a sum of elements of S then $\mathbf{X}^t \in B$. Conversely suppose $\mathbf{X}^t \in B$ and there is an explicit polynomial in the given generators of B which gives $\mathbf{X}^t$. This polynomial must contain a single term which expresses $\mathbf{X}^t$ as a product of elements in $\{\mathbf{X}^s\}_{s \in S}$. Hence, it gives an expression of t as a sum of elements of S. This problem of determining membership of $\mathbf{X}^t$ in B and giving a polynomial in the generators of B which express $\mathbf{X}^t$ if $\mathbf{X}^t$ does lie in B is solved in [16].

1.10 Lemma. *Suppose* $b \in k[\mathbf{X}] \supset S$ *and assume that* b *has an* M^{th} *subductum* b_M *over* S, *i.e.* (1.5) *iterates at least* M *times.*

1.10.a:
$$b = b_M + \sum_{i=0}^{M-1} \gamma_i S^{E_i}$$

---

[8]This includes being able to constructively test for equality.

$$LT\, S^{E_{M-1}} < LT\, S^{E_{M-2}} < \cdots < LT\, S^{E_1} < LT\, S^{E_0}$$

1.10.b:

$$\| \qquad\qquad \| \qquad\qquad\qquad \| \qquad\qquad \|$$

$$LT\, b_{M-1} \qquad LT\, b_{M-2} \qquad\qquad LT\, b_1 \qquad LT\, b_0 = LT\, b$$

and

$$LT\, b_M \; < \; LT\, b_{M-1} \qquad if \qquad b_M \neq 0.$$

1.10.c: *If $0 \neq \mu \in k$ then μb subduces to μb_M.*

1.10.d: *If S lies in a subalgebra B of $k[\mathbf{X}]$: $b \in B$ if and only if $b_M \in B$.*

1.10.e: *The subduction procedure (1.5) always terminates.*

PROOF: Putting together the individual steps $b_{i+1} = b_i - \gamma_i S^{E_i}$ gives $b_M = b_0 - \sum_{i=0}^{M-1} \gamma_i S^{E_i}$. Since $b_0 = b$, we have (1.10.a).

(1.5.d) — together with the fact that b_M occurs — implies:

None of $b_0, \cdots, b_{M-1}$ lie in k and so are not zero.

$LT\, b_{M-1} < LT\, b_{M-2} < \cdots < LT\, b_0 = LT\, b$

$LT\, b_M < LT\, b_{M-1}$ if $b_M \neq 0$.

This gives part (1.10.b).

(1.10.c): Let $c = \mu b$. We subduce c based on the subduction of b and preserve $c_j = \mu b_j$ at each stage. Initially $c_0 = c = \mu b_0$. Suppose we have subduced i times and $c_i = \mu b_i$. Then $b_i \in k$ if and only if $c_i \in k$, for step (1.5.b). Say we get past step (1.5.b). Since $LT\, c_i = LT\, b_i$, both b_i and c_i fare the same at step (1.5.c). If we get past step (1.5.c), and $b_{i+1} = b_i - \gamma_i S^{E_i}$ in step (1.5.e), let $c_{i+1} = c_i - \mu\gamma_i S^{E_i}$ in step (1.5.e). This completes the induction.

(1.10.d) follows from (1.10.a) since $\sum_{i=0}^{M-1} \gamma_i S^{E_i} \in B$.

(1.10.e) follows from (1.10.b) since a term ordering is a well ordering, [13, p. 9, 3.5]. ∎

1.11 Definition. Let B be a subalgebra of $k[\mathbf{X}]$ and $S \subset B$. *S is a SAGBI basis for B* if the lead term of every element of $B \setminus k$ lies in $PP(LT\, S) = LT(PP\, S)$. Since a SAGBI basis generates the subalgebra for which it is a SAGBI basis (1.16.b), we may say: *S is a SAGBI basis*, without specifying a subalgebra, to mean that S is a SAGBI basis for the subalgebra it generates.

One can define *reduced SAGBI bases* by analogy to reduced Gröbner bases. That course is not pursued in this paper.

1.12 Example. Let B be the subalgebra k of $K[\mathbf{X}]$. The condition on elements of $B \setminus k$ is satisfied vacuously since $k \setminus k$ is the empty set. Thus any subset of B, including the empty set, is a SAGBI basis for B. This shows that the empty set is a SAGBI basis (for k).

Note that a subalgebra is a SAGBI basis for itself. Not a particularly useful SAGBI basis, but this shows that every subalgebra has a SAGBI basis. What is desired is a finite SAGBI basis. As will be seen, the existence of a finite SAGBI basis may depend on the term ordering.

1.13 Example. [9] The subalgebra of symmetric polynomials is well known to be generated by elementary symmetric polynomials, [8, p. 204]. A classical proof of this result also shows that the elementary symmetric polynomials form a SAGBI basis for the subalgebra of symmetric polynomials with respect to the lexicographic term ordering.

An amazing phenomenon is that the elementary symmetric polynomials form a SAGBI basis for the subalgebra of symmetric polynomials with respect to every term ordering! We establish this now.[10]

1.14 Theorem. *Suppose that $k[\mathbf{X}]$ has an arbitrary term ordering. The elementary symmetric polynomials form a SAGBI basis for the subalgebra of symmetric polynomials.*

PROOF: A term ordering is a total ordering and so it is at most a matter of renaming variables to be able to assume: $X_1 > X_2 > X_3 > \cdots > X_n$. The first step is to establish:

1.15. If f is a symmetric polynomial then the exponent (vector), $(e_1, e_2, e_3, \cdots, e_n)$ of the lead term of f is *non-increasing*. I. e. if $X_1^{e_1} \cdots X_n^{e_n}$ for $e_1, \cdots, e_n \in \mathbf{N}$ is the lead term of f then: $e_1 \geq e_2 \geq e_3 \geq \cdots \geq e_n$.

Proof of (1.15): Suppose $T_1 = X_1^{d_1} \cdots X_n^{d_n}$ is a term of f with $1 \leq i \leq j \leq n$ where $d_i < d_j$. Let U be the term:

$$X_1^{d_1} X_2^{d_2} \cdots X_{i-1}^{d_{i-1}} X_i^{d_i} X_{i+1}^{d_{i+1}} \cdots X_{j-1}^{d_{j-1}} X_j^{d_i} X_{j+1}^{d_{j+1}} \cdots X_n^{d_n}$$

and let T_2 be the term:

$$X_1^{d_1} X_2^{d_2} \cdots X_{i-1}^{d_{i-1}} X_i^{d_j} X_{i+1}^{d_{i+1}} \cdots X_{j-1}^{d_{j-1}} X_j^{d_i} X_{j+1}^{d_{j+1}} \cdots X_n^{d_n}.$$

Since $X_i > X_j$ the defining properties of a term ordering imply that $X_i^{d_j - d_i} > X_j^{d_j - d_i}$ and also hence that: $T_2 = U X_i^{d_j - d_i} > U X_j^{d_j - d_i} = T_1$. Since f is a symmetric polynomical, T_2 must also be a term of f. Hence T_1 is not a lead term of f and (1.15) is established.

(1.15) implies that the lead terms of the elementary symmetric polynomials are the (usual): $X_1, X_1 X_2, X_1 X_2 X_3, \cdots, X_1 \cdots X_n$. It is an easy exercise/observation that a term with non-increasing exponent vector is a power product of the elements: $X_1, X_1 X_2,$ $X_1 X_2 X_3, \cdots, X_1 \cdots X_n$. In view of (1.15), we have established that the lead term of a symmetric polynomial is a power product of lead terms of elementary symmetric polynomials. ∎

Combining (1.14) with (1.16.b) gives the well known result that the elementary symmetric polynomials generate the subalgebra of symmetric polynomials.

1.16 Proposition. *Suppose $b \in k[\mathbf{X}] \supset S$ and S is a SAGBI basis for a subalgebra B of $k[\mathbf{X}]$. Assume that a subduction of b over S terminated with c as the final subductum.*
1.16.a: *$b \in B$ if and only if $c \in k$.*
1.16.b: *S generates the subalgebra B.*

PROOF: (1.16.a): By (1.10.d), $b \in B$ if and only if $c \in B$. If $c \in k$ then $c \in B$. Conversely suppose $c \notin k$. Since subduction terminated at c, it follows that c does not

---

[9] Thanks to Bernd Sturmfels for this example
[10] Which is why we did not verify the SAGBI basis assertion in (1.13).

have a first subduction over S. (1.5.d) together with $c \notin k$ implies that LT c does not lie in $PP(LT\,S) = LT(PP\,S)$. By (1.11) it follows that $c \notin B$.

(1.16.b): Let b be an element of $B \setminus k$. By part (1.16.a), a terminating subduction for b must terminate at an element of k. Suppose this element of k is b_M, the Mth subductum of b over S. By (1.10.a), $b = b_M + \sum_{i=0}^{M-1} \gamma_i S^{E_i}$. Thus S generates B. ∎

1.17 Example. Suppose S is a non-empty set of terms.[11] Let b be an element in the subalgebra generated by S where b does not lie in k. LT b is an S power product and so S equals LT S. Thus S is a SAGBI basis (1.11), for the subalgebra it generates. In particular, $\{X_1, \cdots, X_n\}$ is a SAGBI basis for $k[\mathbf{X}]$.

The previous example shows that for a monoid subalgebra of $k[\mathbf{X}]$, with any term ordering, a generating set for the monoid is a SAGBI basis for the subalgebra. (1.17) combined with (4.6) shows that the theory of SAGBI bases is algorithmic for monoid subalgebras generated by finitely generated submonoids. In this case, SAGBI bases may be used to compute and prove results for monoid algebras.

1.18 Example. Let $k[X]$, the polynomial ring in one variable, have its usual, unique term ordering.[12] Suppose S is the subset $\{X^2 + X, X^2\}$ of $k[X]$. Since $X = (X^2 + X) - X^2$, the subalgebra of $k[X]$ generated by S is $k[X]$ itself. X^2 is the only lead term of elements of S and X is not a power product of X^2. Thus S generated $k[X]$ but S is not a SAGBI basis for $k[X]$.

1.19 Proposition. *Let B be a subalgebra of $k[\mathbf{X}]$ generated by a single element. A set $S \subset B$ is SAGBI basis for B if and only if S contains an element which generates B. Any element which generates B is a singleton SAGBI basis for B.*

PROOF: Left to reader. ∎

1.20 Example of a finitely generated graded subalgebra which has no finite SAGBI basis. Let B be the subalgebra of $k[X, Y]$ (finitely) generated by $X + Y, XY, XY^2$. Since B is generated by homogeneous elements, B is a graded subalgebra of $k[X, Y]$. We shall show that B does not have a finite SAGBI basis with any term ordering. This first step in this direction is to produce a useful family of elements in B and a useful family of elements which are not in B.

IN B: XY and XY^2 lie in B. $XY^n = (X + Y)XY^{n-1} - (XY)XY^{n-2}$ provides the induction step showing that $XY^n \in B$ for all $1 \le n \in \mathbf{N}$.

NOT IN B: For $j \ge 1$, B contains no elements having Y^j as a homogeneous component, and hence no elements having non-zero scalar times Y^j as a homogeneous component. Explanation: since B is a graded subalgebra of $k[X, Y]$, for any element of B, each of its homogeneous components lie in B. Hence B would contain Y^j. This cannot happen because all elements of B can be expressed in the form $h(X + Y, XY, XY^2)$ as h ranges over polynomials in three variables. Suppose there were an h for which $h(X + Y, XY, XY^2) = Y^j$. Setting X to zero gives $h(Y, 0, 0) = Y^j$. Setting Y to zero in $h(X + Y, XY, XY^2) = Y^j$ gives $h(X, 0, 0) = 0$. But since $h(Y, 0, 0) = Y^j$ it follows that $h(X, 0, 0) = X^j$. Thus $X^j = 0$, a contradiction.

---

[11] *terms*, not general polynomials!

[12] Where $X^i < X^j$ if and only if $i < j$.

For term orderings on $k[X, Y]$, either $X > Y$ or vice versa. Suppose $k[X, Y]$ has been assigned a term ordering with $X > Y$. By *IN B*, $S = \{X + Y, XY, XY^2, XY^3, \cdots\} \subset B$. We show that S is a SAGBI basis for B and that B has no finite SAGBI basis.

BASIS: For $j \geq 1$, B contains no elements having Y^j as a lead term. Explanation: suppose f is a polynomial with lead term Y^j. In the term ordering, Y^j is the smallest term of degree j. Hence the degree j component of f must be a non-zero scalar times Y^j. By *NOT IN B*, f cannot lie in B. Hence the lead term of any non-constant element b of B is of the form $X^i Y^j$ with $i \geq 1$. Thus $\mathrm{LT}\, b = X^{i-1} X Y^j = \mathrm{LT}(X+Y)^{i-1}\, \mathrm{LT}\, XY^j$ and S is a SAGBI basis for B.

NOT A BASIS: Suppose B has a *finite* SAGBI basis T. Choose m large enough so that no element of T has lead term XY^m. Since T is a SAGBI basis for B, XY^m is a power product of lead terms of elements of T. Since X occurs to the first power, it follows that T has an element with lead term of the form XY^i with $0 \leq i$ and an element with lead term Y^i with $1 \leq j$. This contradicts *NOT IN B*.

Suppose $k[X, Y]$ has been assigned a term ordering with $Y > X$. Note that B is also generated by $X + Y, XY, X^2Y$ since $(X + Y)XY = X^2Y + XY^2$. With X and Y interchanged, the same reasoning as above shows that $\{X + Y, XY, X^2Y, X^3Y, \cdots\}$ is a SAGBI basis for B and B has no finite SAGBI basis. ∎

Note that in the previous example, $k[X, Y]$ is integral over B. Both X and Y satisfy the integral equation: $Z^2 - (X + Y)Z + XY$.

2. SAGBI Basis Test

We introduce convenient terminology concerning *representations* of sums, as opposed to the actual values of the sums. A sum appears in quotes when we are concerned with its representation.

2.1 Definition. Given a sequence $(f_1, \cdots, f_M)$ with f_i's in $k[\mathbf{X}]$, *where not all the f_i's are zero*, let H be the maximal lead term which occurs among the non-zero f_i's. H is called the *height* of the sequence. The number of non-zero f_i's with $\mathrm{LT}\, f_i = H$ is the *breadth* of the sequence. The *sum* of the sequence is, of course: $\sum_1^M f_i$. For a sequence of polynomials, all of which are zero, the height of the sequence is not defined but the breadth and sum of the sequence are zero. A summation in quotes, "$\sum_1^M f_i$", indicates the sum of the sequence: $(f_1, \cdots, f_M)$.[13] Since "$\sum_1^M f_i$" has the *abiding memory* of $(f_1, \cdots, f_M)$, the *breadth* of "$\sum_1^M f_i$" means the breadth of $(f_1, \cdots, f_M)$. Similarly for the *height* of "$\sum_1^M f_i$" when not all the f_i's are zero.

2.2 Remarks. Of course it may happen that "$\sum_1^M f_i$" $=$ "$\sum_1^N g_i$" while "$\sum_1^M f_i$" and "$\sum_1^M g_i$" have different height or breadth. If "$\sum_1^M f_i$" $\neq 0$ then $(f_1, \cdots, f_M)$ must have at least one non-zero polynomial and the height of "$\sum_1^M f_i$" is defined. The height of a sequence is defined if and only if the breadth is non-zero.

Height has the following easy properties. In the following, f_i's and g_i's lie in $k[\mathbf{X}]$. Each item is followed by its justification.

---

[13] Thus $f = $ "$\sum_1^N f_1$" makes sense.

2.2.a: If "$\sum_1^M f_i$" $\neq 0$ then LT("$\sum_1^M f_i$") is less than or equal to the height of "$\sum_1^M f_i$". (By (1.4.a).)

2.2.b: If the breadth of "$\sum_1^M f_i$" is one then "$\sum_1^M f_i$" $\neq 0$ and LT("$\sum_1^M f_i$") equals the height of "$\sum_1^M f_i$". (By (1.4.b).)

2.2.c: If either "$\sum_1^M f_i$" or "$\sum_1^N g_j$" has a height[14] then "$\sum_1^M f_i + \sum_1^N g_j$" has a height which is no greater than the heights of "$\sum_1^M f_i$" or "$\sum_1^N g_j$". (From the definition.)

2.2.d: If "$\sum_1^M f_i$" and "$\sum_1^N g_j$" have height H_f and H_g resp. then "$\sum_{i=1}^M \sum_{j=1}^N f_i g_j$" has height $H_f H_g$. (By (1.4.c).)

2.2.e: Suppose $b \in k[\mathbf{X}] \supset S$ and $b = b_0$ subduces to b_M where $M \geq 1$. With respect to the γ_i's and E_i's which occur in this subduction: "$b_M + \sum_{i=0}^{M-1} \gamma_i S^{E_i}$" is a breadth one sum of height LT(S^{E_0}). (By (1.10.b).)

2.3 Proposition. *Let $S \subset k[\mathbf{X}]$ and let B be the subalgebra of $k[\mathbf{X}]$ generated by S. The following conditions are equivalent:*

2.3.a: *S is a SAGBI basis.*

2.3.b: *Every subduction over S of elements in $B \setminus k$ terminates at an element of k.*

2.3.c: *Each element in $B \setminus k$ has at least one subduction over S which terminates at an element of k.*

2.3.d: *Each element b in $B \setminus k$ can be expressed as a sum of breadth one of the form:* "$\sum_i \gamma_i S^{E_i}$" *with $\gamma_i \in k$ and E_i's exponent functions on B.*

PROOF: (2.3.a) implies (2.3.b) by (1.16.a). Clearly (2.3.b) implies (2.3.c).

(2.3.c) implies (2.3.d): Let b be an element of $B \setminus k$ and suppose $b_M \in k$ is the final subductum of a subduction over S of $b = b_0$. Since $b = b_0 \notin k$, and $b_M \in k$ it follows that the subduction iterates at least once. Applying (1.10.a), with respect to the γ_i's and E_i's which occur in this subduction of $b = b_0$, $b = b_M + \sum_{i=0}^{M-1} \gamma_i S^{E_i}$. Since $b_M \in k$ let $\lambda_M = b_M$ and let E_M be the zero exponent function (1.1). Then $b =$ "$\sum_i \gamma_i S^{E_i}$". By (2.2.e), this is a breadth one sum.

(2.3.d) implies (2.3.a): Suppose that each non-zero element of the subalgebra generated by S can be expressed as a breadth one sum of the form: "$\sum_i \gamma_i S^{E_i}$" with $\gamma_i \in k$ and the E_i's exponent functions. By (2.2.b): LT("$\sum_i \gamma_i S^{E_i}$") equals the height of "$\sum_i \gamma_i S^{E_i}$" which is the maximal LT S^{E_i} among the i's with $\gamma_i \neq 0$. In particular, each non-zero element of the subalgebra generated by S has a lead term which is the lead term of an S power product. Thus S is a SAGBI basis (1.11). ∎

Buchberger theory has its S-polynomials. The SAGBI analog is the *tête-a-tête*. Where $S(f,g)$, the S-polynomial of two polynomials f and g, relates Buchberger reduction by f to reduction by g, the *tête-a-tête* of two exponent functions E and F relates subduction by E to subduction by F. In Buchberger theory, $S(f,g)$ is essentially the same as $S(g,f)$,[15] and one only bothers with one of the two. The same is true of *tête-a-têtes*.

---

[14] I.e. either some f_i or some g_j is non-zero

[15] In most treatments of Buchberger theory: $S(g,f) = \lambda S(g,f)$, for $0 \neq \lambda \in k$. In some treatments of Buchberger theory, λ always equals -1.

2.4 Definition. Suppose E and F are exponent functions on a subalgebra R of $k[\mathbf{X}]$. The pair (E, F) is a R *tête-a-tête*[16] if LT $R^E =$ LT R^F.[17] This common value, LT $R^E =$ LT R^F, is called the *height* of the tête-a-tête. For a tête-a-tête (E, F) there is a non-zero element $\rho = \rho(E, F) \in k$ where R^E and ρR^F have the same lead monomial. For such $0 \neq \rho \in k$: let $T(E, F)$ denote $R^E - \rho R^F$. Note that $\rho(F, E) = 1/\rho(E, F)$. Hence: $T(E, F) = -\rho T(F, E)$.

Obviously (E, F) is a tête-a-tête if and only if (F, E) is a tête-a-tête. If so, $T(E, F) = \lambda T(F, E)$, for $0 \neq \lambda \in k$. $T(E, F) = 0$ if and only if R^E equals a scalar times R^F. Suppose (E, F) and (E', F') are tête-a-têtes and $m \in \mathbf{N}$. $(E + E', F + F')$ and (mE, mF) are tête-a-têtes.

2.5 Example. Suppose $S = \{X, X^2\} \subset k[X]$. Let E and F be exponent functions on the subalgebra generated by S which take the value zero except for: $E(X) = 2$ and $F(X^2) = 1$. Then $S^E = (X)^2 (X^2)^0$ and $S^F = (X)^0 (X^2)^1$. Since $S^E = X^2 = S^F$ they have the same lead term and (E, F) is a tête-a-tête. Moreover, $T(E, F) = 0$, although $E \neq F$.

The next major result is the SAGBI basis test (2.8) which is the SAGBI analog to Buchberger's Gröbner basis test. In both the SAGBI and Buchberger theory, the basis test is the key to the construction of bases. There are generally an infinite number of tête-a-têtes but the set of tête-a-têtes may be finitely generated and it is only necessary to use a generating set for the SAGBI basis test. Algorithmically finding a generating set for tête-a-têtes is discussed at (2.20). Here is what it means for a set to generate the tête-a-têtes.

2.6 Definition. Suppose S is a subset of $k[\mathbf{X}]$. A set T *generates* the S tête-a-têtes if T is a set of S tête-a-têtes and for any S tête-a-tête (E, F) there exist S tête-a-têtes (L_i, R_i)'s and m_i's in $\mathbf{N}$ satisfying:

$$\text{for each } i \text{ either} \quad (L_i, R_i) \in T \quad \text{or} \quad (R_i, L_i) \in T$$

2.7
$$E = \sum_i m_i L_i \quad \text{and} \quad F = \sum_i m_i R_i$$

The sets S and T are *not* assumed to be finite.

2.8 Theorem. *Let S be a subset of $k[\mathbf{X}]$ and let T be a set which generates the S tête-a-têtes. The following are equivalent:*
2.8.a: *S is a SAGBI basis.*
2.8.b: *For each S tête-a-tête $(L, R) \in T$, with $T(L, R) \neq 0$, there exist exponent functions G_j's on S and λ_j's in k satisfying:*

$$T(L, R) = \text{``} \sum_j \lambda_j S^{G_j} \text{''}$$

the height of $\text{``} \sum_j \lambda_j S^{G_j} \text{''}$ *is less than the height of (L, R)*

---

[16] The name *tête-a-tête* has been chosen because R^E and R^F have the same *head* term.
[17] We may say *tête-a-tête*, dropping the R. If S is a subset of R containing the support of both E and F we may say (E, F) is an S *tête-a-tête*. The *set of S tête-a-têtes* consists of the set of all tête-a-têtes with support in S on the subalgebra of $k[\mathbf{X}]$ generated by S. This is essentially the same as the set of all tête-a-têtes on $k[\mathbf{X}]$ with support in S.

2.8.c: *For each S tête-a-tête $(L, R) \in T$, $T(L, R)$ has at least one subduction over S which terminates at an element of k.*

2.8.d: *For each S tête-a-tête (L, R), every subduction of $T(L, R)$ over S terminates at an element of k.*

The proof is modeled on the proof in [17] and [13] of Buchberger's Gröbner basis test. The proof uses a technical lemma (2.13) which has been placed after the theorem.

PROOF: Let B be the subalgebra of $k[\mathbf{X}]$ generated by S. If S is a SAGBI basis then every element of B, including elements of the form $S^L - \rho S^R$, subduce to elements of k over S, (1.16.a). Thus (2.8.a) implies (2.8.d). Obviously (2.8.d) implies (2.8.c). Suppose $(L, R) \in T$ and $T(L, R) \neq 0$. Set $b_0 = T(L, R)$. By (2.8.c), b_0 has a (say, M-step) subduction to an element $b_M \in k$. Applying (1.10.a), with respect to the γ_i's and E_i's which occur in this subduction of $b_0 : b_0 = b_m + \sum_{i=0}^{M-1} \gamma_i S^{E_i}$. Let E_M be the exponent function 0, (1.1). Then $b_0 = \sum_{i=0}^{M} \gamma_i S^{E_i}$. By (1.10.b), " $\sum_{i=0}^{M} \gamma_i S^{E_i}$ " has height less than or equal to b_0. Thus b_0 has a sum of the desired form and (2.8.c) implies (2.8.b).

It remains to show that (2.8.b) implies (2.8.a). Suppose (2.8.b) holds and let d be any non-zero element of B. Since S generates B as an algebra, there are $\omega_i \in k$ and exponent functions F_i's on B where d equals:

$$2.9 \qquad\qquad `` \sum_{i=0}^{M} \omega_i S^{F_i} "$$

Among the possible sums (2.9), choose F_i's and non-zero ω_i's so that the height of (2.9) is as small as possible and secondarily the breadth of (2.9) is as small as possible. We shall show that the breadth of this doubly minimal sum equals 1. By (2.3) it follows that S is a SAGBI basis. Suppose the breadth of the doubly minimal sum is at least 2. This will be shown to be a contradiction. By renumbering, we can assume that the first two summands of (2.9) have lead term equal to the height of (2.9). Thus the height of (2.9) equals LT $S^{F_1} =$ LT S^{F_2}. By assumption $\omega_1 \neq 0 \neq \omega_2$.

(F_1, F_2) is an S tête-a-tête because LT $S^{F_1} =$ LT^{F_2}. Let ρ be the non-zero element of k where $S^{F_1} - \rho S^{F_2}$ equals $T(F_1, F_2)$. Thus:

$$* \qquad\qquad \omega_1 S^{F_1} + \omega_2 S^{F_2} = \omega_1(S^{F_1} - \rho S^{F_2}) + (\omega_1 \rho + \omega_2) S^{F_2}$$

Suppose $T(F_1, F_2) \neq 0$. (The next case to be considered will be where $T(F_1, F_2) = 0$.) Since (2.8.b) is assumed to hold, (2.13) applies and $T(F_1, F_2)$ can be expressed as a sum of height less than the height of (F_1, F_2) of the form: " $\sum_h \gamma_j S^{E_j}$ " with $\gamma_j \in k$ and E_j's, exponent functions. In this case:

$$\omega_1 S^{F_1} + \omega_2 S^{F_2} = \sum_h \omega_1 \gamma_j S^{E_j} + (\omega_1 \rho + \omega_2) S^{F_2}$$

The (2.9) can be rewritten:

$$2.10 \qquad\qquad ``(\omega_1 \rho + \omega_2) S^{F_2} + \sum_{i=2}^{M} \omega_i S^{F_1} + \sum_h \omega_1 \gamma_j S^{E_j} "$$

The height of " $\sum_h \gamma_j S^{E_j}$ " is less than the height of (F_1, F_2) which equals LT S^{F_2} which equals the height of (2.9). Hence " $\sum_h \omega_1 \gamma_j S^{E_j}$ " does not contribute to the height or breadth of (2.10).

What can be said about the height and breadth of (2.10)? Three cases arise:

1. If $(\omega_1 \rho + \omega_2)$ is not zero then the presence of the monomial $(\omega_1 \rho + \omega_2) S^{F_2}$ shows that (2.10) has height LT S^{F_2} the same as (2.9). The absence of $\omega_1 S^{F_1}$ shows that (2.10) has breadth one less than (2.9). This contradicts the minimality of the breadth of (2.9).

2. If $(\omega_1 \rho + \omega_2)$ is zero, (2.10) equals

2.11
$$\text{``} \sum_{i=2}^{M} \omega_i S^{F_i} + \sum_{h} \omega_1 \gamma_j S^{E_j} \text{''}$$

and there are two further cases to consider:

2A. (2.9) has breadth three or greater. In this case a later (than $i = 2$) monomial of (2.9) has lead term equal to LT S^{F_2}. In this case (2.11) has height LT S^{F_2} the same as (2.9). The absence of $\omega_1 S^{F_1}$ and $\omega_2 S^{F_2}$ shows that (2.10) has breadth two less than (2.9). This contradicts the minimality of the breadth of (2.9).

2B. (2.9) has breadth precisely two. In this case (2.11) has height less than the height of (2.9), contradicting the minimality of the height of (2.9).

Finally the case where $T(F_1, F_2) = 0$. In this case:

$$\omega_1 S^{F_1} + \omega_2 S^{F_2} = (\omega_1 \rho + \omega_2) S^{F_2}$$

and (2.9) can be rewritten:

2.12
$$\text{``} (\omega_1 \rho + \omega_2) S^{F_2} + \sum_{i=2}^{M} \omega_i S^{F_i} \text{''}$$

The rest of the analysis proceeds similarly to when $T(F_1, F_2) \neq 0$. Three cases result. All contradictions. ∎

Here is the technical result used above which is key to showing that it is only to use a generating set for the tête-a-têtes in the basis test. Again the sets S and T are *not* assumed to be finite.

2.13 Lemma. *Suppose S is a subset of $k[X]$ and suppose T generates the S tête-a-têtes. Suppose that for each $(L, R) \in T$ with $T(L, R) \neq 0$, there exist exponent functions G_j's on S and λ_j's in k satisfying:*

$$T(L, R) = \text{``} \sum_j \lambda_j S^{G_j} \text{''},$$

2.14
$$\text{the height of ``} \sum_j \lambda_j S^{G_i} \text{'' is less than the height of } (L, R).$$

(The G_j's and λ_j's depend on (L, R).) Then for every S tête-a-tête (E, F) with $T(E, F) \neq 0$, there exist exponent functions H_j's on S and μ_j's in k satisfying:

$$T(E, F) = \text{``} \sum_j \mu_j S^{H_j} \text{''},$$

2.15
$$\text{the height of ``} \sum_i \mu_j S^{H_i} \text{'' is less than the height of } (E, F).$$

(The H_j's and μ_j's depend on (E,F).)

PROOF: As noted at (2.4): $T(L,R)$ equals a scalar times $T(R,L)$. Thus the fact that (2.14) holds for (L,R) with $(L,R) \in T$ implies that (2.14) holds for (R,L) with $(L,R) \in T$. This will be used several lines down where we assume (2.14) holds for (L_i, R_i) where either $(L_i, R_i) \in T$ or $(R_i, L_i) \in T$. Suppose (E,F) is an S tête-a-tête. Since T generates the S tête-a-têtes, there exist S tête-a-têtes (L_i, R_i)'s and m_i's in $\mathbf{N}$ where (2.7) is satisfied. For each (L_i, R_i) let ρ_i be the unique *non-zero* element of k where S^{L_i} has the same lead monomial as $\rho_i S^{R_i}$. If $T(L_i, R_i) = 0$, then: $S^{L_i} = \rho_i S^{R_i}$. For i where $T(L_i, R_i) \neq 0$, taking into account that the G_j's and λ_j's at (2.14) depend on (L_i, R_i):

$$S^{L_i} = \rho_i S^{R_i} + \sum_j \lambda_{i,j} S^{G_{i,j}}$$

By (2.7):

2.16 $$S^E = \prod_i (S^{L_i})^{m_i} = \left(\prod_{\substack{i \text{ where} \\ T(L_i,R_i)=0}} \left(\rho_i S^{R_i} \right)^{m_i} \right)$$
$$\times \left(\prod_{\substack{i \text{ where} \\ T(L_i,R_i)\neq 0}} \left(\rho_i S^{R_i} + \sum_j \lambda_{i,j} S^{G_{i,j}} \right)^{m_i} \right)$$

Suppose the right hand product is non-empty, i.e. there are i where $T(L_i, R_i)$ is non-zero. In the right hand product: by definition, the height of the tête-a-tête (L_i, R_i) equals LT S^{R_i}. By hypothesis, the height of (L_i, R_i) is greater than the height of the sum "$\sum_j \lambda_{i,j} S^{G_{i,j}}$". Thus LT $\rho_i S^{R_i}$ is the largest term in each of the productands $(\rho_i S^{R_i} + \sum_j \lambda_{i,j} S^{G_{i,j}})^{m_i}$. Because the term ordering is multiplicative, the product of the largest terms is the largest term of the product. Thus the largest term of (2.16) is computed:

$$\left(\prod_{\substack{i \text{ where} \\ T(L_i,R_i)=0}} \mathrm{LT} \left(\rho_i S^{R_i} \right)^{m_i} \right) \left(\prod_{\substack{i \text{ where} \\ T(L_i,R_i)\neq 0}} \mathrm{LT} \left(\rho_i S^{R_i} + \sum_j \lambda_{i,j} S^{G_{i,j}} \right)^{m_i} \right)$$
$$= \left(\prod_{\substack{i \text{ where} \\ T(L_i,R_i)=0}} \mathrm{LT} \left(\rho_i S^{R_i} \right)^{m_i} \right) \left(\prod_{\substack{i \text{ where} \\ T(L_i,R_i)\neq 0}} \mathrm{LT}(\rho_i S^{R_i})^{m_i} \right)$$
$$= \prod_i \mathrm{LT} \left(\rho_i S^{R_i} \right)^{m_i}$$
$$= \mathrm{LT} \prod_i \left(\rho_i S^{R_i} \right)^{m_i}$$

Thus (2.16), and hence S^E, equals:

2.17 $$\left(\prod_i (\rho_i S^{R_i})^{m_i} \right) + \sum_h \mu_h S^{H_h}$$

where H_h's are exponent functions which are sums of R_i's and $G_{i,j}$'s and the height of "$\sum_h \mu_h S^{H_h}$" is less than $\mathrm{LT} \prod_i (\rho_i S^{R_i})^{m_i}$. (NOTE, none of the ρ_i's are zero so $\prod_i (\rho_i S^{R_i})^{m_i}$) is not zero.)

Let $\rho = \prod_i \rho_i^{m_i}$ and note $\prod_i (S^{R_i})^{m_i} = S^F$ since $F = \sum_i m_i R_i$. Thus the height of "$\sum_h \mu_h S^{H_h}$" is less than $\mathrm{LT}\, S^F$ which is the height of (E, F). Using ρ and this expression for S^F to simplify "$(\prod_i (\rho_i S^{R_i})^{m_i})$" at (2.17) gives for (2.17):

2.18
$$\rho S^F + \sum_h \mu_h S^{H_h}$$

Putting together (2.16) through (2.18):

$$S^E = \rho S^F + \sum_h \mu_h S^{H_h}$$

2.19
the height of "$\sum_h \mu_h S^{H_h}$" is less than the height of (E, F)

It only remains to show that ρ is the unique scalar with: $S^E - \rho S^F = T(E, F)$. (2.19) shows that $\mathrm{LT}(S^E - \rho S^F)$ is less than or equal to the height of "$\sum_h \mu_h S^{H_h}$" which is less than the height of (E, F) which equals $\mathrm{LT}\, S^E$ and $\mathrm{LT}\, S^F$. Thus the lead monomials of S^E and ρS^F must cancel in $S^E - \rho S^F$ and ρ must be the unique non-zero element of k with $S^E - \rho S^F = T(E, F)$. (2.19) shows that $T(S, F)$ can be expressed as a sum in the desired form (NOTE, it may happen that the $T(E, F) = 0$ but can be expressed as a sum "$\sum_h \mu_h S^{H_h}$" of height less than (E, F).)

Finally, the case where all $T(L_i, R_i)$'s are zero, i. e. right hand product at (2.16) is empty. With similar reasoning and notation to the previous case: $S^E = \prod_i (\rho_i S^{R_i})^{m_i} = (\prod_i \rho_i^{m_i})(\prod_i (S^{R_i})^{m_i}) = \rho S^F$ and ρ is the unique non-zero element of k where $S^E - \rho S^F = T(E, F)$. Thus $T(E, F) = 0$ and there is nothing to prove. ∎

As in classical Buchberger theory, the test theorem is the key to constructing SAGBI bases. Namely, if a set S is not a SAGBI basis there will be at least one S tête-a-tête (L, R) where $T(L, R)$ does not subduce to an element of k. These final subducts of elements not in k are what to add to the set to get closer to SAGBI basis. The details appear in the next section, especially (3.1) and (3.5).

2.20. GENERATING TÊTE-A-TÊTES: The following discussion is about finding generators for S tête-a-têtes algorithmically. Finding tête-a-têtes and generators for tête-a-têtes for a finite set $S \subset k[\mathbf{X}]$ can be done by finding non-negative integer solutions and bases for non-negative integer solutions to equations with integer coefficients. References can be found in the earlier remark (1.9). The reduction from tête-a-têtes to equations with integer coefficients relies on the correspondence between (exponent) vectors in $\mathbf{N}^n$ and terms in $k[\mathbf{X}] = k[X_1, \cdots, X_n]$. If $\mathbf{e} = (e_1, \cdots, e_n) \in \mathbf{N}^n$, $\mathbf{e}$ is the exponent vector of the term $X_1^{e_1} X_2^{e_2} \cdots X_n^{e_n} \in k[\mathbf{X}]$. We freely use that taking power products of terms corresponds to forming linear combinations with non-negative coefficients of their exponent vectors.

Let S be a finite subset of $k[\mathbf{X}] = k[X_1, \cdots, X]$. For each $s \in S$ let $\mathbf{e}_s \in \mathbf{N}^n$ be the exponent vector of the lead term of s, let $\{Y_s\}_{s \in S}$ and $\{Z_s\}_{s \in S}$ be variables and consider the system of n linear equations given by:

2.21
$$\sum_{s \in S} \mathbf{e}_s Y_s - \sum_{s \in S} \mathbf{e}_s Z_s = 0$$

where 0 is the zero vector in $\mathbf{N}^n$. $\mathbf{y} = (y_s)_{s \in S}$ and $\mathbf{z} = (z_s)_{s \in S}$ are vectors which give a solution to the system (2.21) if and only if $\sum_{s \in S} \mathbf{e}_s y_s = \sum_{s \in S} \mathbf{e}_s z_s$. If $\mathbf{y}$ and $\mathbf{z}$ are vectors of non-negative integers, let $L_\mathbf{y}$ and $R_\mathbf{z}$ be the exponent functions on R which take the value zero except for:

2.22 $$L_\mathbf{y}(s) = y_s, \qquad R_\mathbf{z}(s) = z_s \qquad \text{for } s \in S$$

Then $(L_\mathbf{y}, R_\mathbf{z})$ is an S tête-a-tête if and only if $\mathbf{y} = (y_s)_{s \in S}$ and $\mathbf{z} = (z_s)_{s \in S}$ are vectors of non-negative integers which give a solution to the system of linear equations. This shows that finding tête-a-têtes is equivalent to finding solutions of (2.21) in non-negative integers.

Let U be the set of vectors $(\mathbf{y}, \mathbf{z})$ where $\mathbf{y} = (y_s)_{s \in S}$ and $\mathbf{z} = (z_s)_{s \in S}$ are vectors of non-negative integers which give a solution to the system (2.21). Under componentwise addition, U is a monoid. Also, if $(\mathbf{y}, \mathbf{z}) \in U$ then $(\mathbf{z}, \mathbf{y}) \in U$.

2.23 Definition. *A set V flip generates U if $V \subset U$ and for any $(\mathbf{y}, \mathbf{z}) \in U$ there exist $(\mathbf{y}_i, \mathbf{z}_i)$'s in U and m_i's in $\mathbf{N}$ satisfying:*

2.24 $$\text{for each } i \text{ either: } (\mathbf{y}_i, \mathbf{z}_i) \in V \text{ or } (\mathbf{z}_i, \mathbf{y}_i) \in V \text{ and } (\mathbf{y}, \mathbf{z}) = \sum_i m_i (\mathbf{y}_i, \mathbf{z}_i)$$

If V flip generates U then the tête-a-têtes arising from V via (2.22) generate the S tête-a-têtes.

2.25 Definition. *A set V (monoid) generates U if $V \subset U$ and for any $(\mathbf{y}, \mathbf{z}) \in U$ there exist $(\mathbf{y}_i, \mathbf{z}_i)$'s in V and m_i's in $\mathbf{N}$ satisfying: $(\mathbf{y}, \mathbf{z}) = \sum_i m_i (\mathbf{y}_i, \mathbf{z}_i)$.*

Obviously, if V *monoid generates* U then V *flip generates* U and gives a generating set for the S tête-a-têtes. [10] and [3] present algorithms for finding monoid generating sets of solutions to the equations with integer coefficients. Combined with (2.6) - (2.25) they give an algorithm for finding generating sets for S tête-a-têtes when S is finite. Let us illustrate the ideas above.

2.26 Example. Suppose $S = \{X^3 + X, X^2\} \subset k[X]$. Let $s \equiv X^3 + X$ and $t \equiv X^2$. The lead terms of elements of S are X^3, X^2. $\mathbf{e}_s = 3$ and $\mathbf{e}_t = 2$ as vector in $\mathbf{N}^1$. The system (2.21) becomes a single equation:

$$3Y_s + 2Y_t - 3Z_s - 2Z_t = 0$$

Writing solution vectors (y_s, y_t, z_s, z_t), a non-negative monoid generating set for this equation is $\{(1,0,1,0), (2,0,0,3), (0,3,2,0), (0,1,0,1)\}$. This may be varified directly as an exercise. From this monoid generating set, choose the flip genereting set $\{(1,0,1,0), (2,0,0,3), (0,1,0,1)\}$. $(1,0,1,0)$ yields the trivial tête-a-tête LT s = LT s. $(0,1,0,1)$ yields the trivial tête-a-tête LT t = LT t. $(2,0,0,3)$ yields the non-trivial tête-a-tête LT s^2 = LT t^3. These three tête-a-têtes generate the S tête-a-têtes.

3. SAGBI basis construction

This section presents SAGBI basis construction. One begins with a generating set for a subalgebra and wishes to find a SAGBI basis for the subalgebra. As in standard Buchberger theory, the test theorem is the key to proving the construction procedure gives a SAGBI basis. The construction we present starts with a possibly infinite set G and leads to a SAGBI basis for the subalgebra B, generated by G. As per example (1.20), even when G is finite, B may not have a finite SAGBI basis. If G is finite and B does have a finite SAGBI basis, our SAGBI basis construction really terminates in a finite number of steps and is algorithmic. This will be explained below. In general, the following is not algorithmic.

To emphasize the underlying principles, optimizations are excluded.

3.1. SAGBI BASIS CONSTRUCTION: Suppose G is a (possibly infinite) subset of $k[\mathbf{X}]$.

Initialize: Set $G_0 = G$.

Inductive step; j^{th} stage with G_j, to $j+1$ stage with G_{j+1}:
1. Let T_j be a set which generates the G_j tête-a-têtes.
2. For each $(L, R) \in T_j$ let $f(L, R) \in k[\mathbf{X}]$ be a final subductum of $T(L, R)$ over G_j. (The final subductum from any subduction of $T(L, R)$ over G_j will do.) Let F_j be the subset of $k[\mathbf{X}]$ consisting of the $f(L, R)$'s which do not lie in k.
3. Let $G_{j+1} = F_j \cup G_j$

Finalize: Set $G_\infty = \bigcup G_j$.

3.2 Remark. Since $G_{i+1} = F_i \cup G_i$ we have $G_0 \subset G_1 \subset G_2 \subset \cdots$.

3.3 Remark. Let B be the subalgebra of $k[\mathbf{X}]$ generated by G. Notice that each G_i and hence G_∞ lie in B.

3.4 Remark. Suppose $G_{j+1} = G_j$. The T_j may be used for T_{j+1}. Using the same final subductums $f(L, R)$ as used at the jth stage, leads to $F_{j+1} = F_j$. Hence $G_{j+2} = F_{j+1} \cup G_{j+1} = F_j \cup G_j = G_{j+1} = G_j$. By iteration: if $G_j = G_{j+1}$ then $G_j = G_{j+1} = G_{j+2} = \cdots$ and $G_j = G_\infty$. In this case, (3.1) really stops at the jth stage.

3.5 SAGBI basis construction theorem. *Suppose G is a (possibly infinite) subset of $k[\mathbf{X}]$, B is the subalgebra of $k[\mathbf{X}]$ generated by G. Using the notation of (3.1):*
   a. *G_∞ is a SAGBI basis for B.*
   b. *G_j is a SAGBI basis for B if and only if $G_j = G_{j+1}$.*

PROOF: a. uses the equivalence of (2.8.a) and (2.8.c) in the basis test theorem (2.8). First we must find a generating set for the G_∞ tête-a-têtes. Let $T_\infty = \bigcup T_j$. Say $(L, R) \in T_i$. Then L and R are exponent functions (1.1) on B with support in G_i and hence with support in G_∞. Thus (L, R) is a G_∞ tête-a-tête. Now to show that T_∞ generates the G_∞ tête-a-têtes. Say (E, F) is a G_∞ tête-a-tête. Since exponent functions have finite support, (1.1), there is a finite subset $H \subset G_\infty$ which contains the support of both E and F. Since H is finite, there is a value j where $H \subset G_j$. Then $(E; F)$ is a G_j tête-a-tête. Since T_j generates the G_j tête-a-têtes, by (2.7) there exist G_j tête-a-têtes (L_i, R_i)'s and m_i's in $\mathbf{N}$ satisfying:

$$\text{for each } i \text{ either} \quad (L_i, R_i) \in T_j \quad \text{or} \quad (R_i, L_i) \in T_j$$
$$E = \sum_i m_i L_i \quad \text{and} \quad F = \sum_i m_i R_i$$

Thus T_∞ generates the G_∞ tête-a-têtes.

Say $(L, R) \in T_\infty$ where $(L, R) \in T_j$. Let $f(L, R)$ be the final subductum of $T(L, R)$ over G_j used in step j of (3.1). If $f(L, R) \notin k$ then $f(L, R) \in G_{j+1}$ as described in (3.1). Thus, in one more step, $T(L, R)$ subduces to zero *over* G_{j+1}. We have shown that for every $(L, R) \in T_\infty$, $T(L, R)$ has at least one subductum over G_∞ which terminates at an element of k. By the equivalence of (2.8.a) and (2.8.c) in the basis test theorem (2.8), it follows that G_∞ is a SAGBI basis, for B.

b. If $G_j = G_{j+1}$, then $G_j = G_\infty$ by (3.2). By (a), G_j is a SAGBI basis for B. Conversely suppose that G_j is a SAGBI basis. By the equivalence of (2.8.d) and (2.8.a) in the basis test theorem (2.8), it follows that for every G_j tête-a-tête (L, R) every final subductum of $T(L, R)$ lies in k. Hence, F_j is empty and $G_{j+1} = G_j$. ∎

Note that the proof of (a) gives a set T_∞ which generates the G_∞ tête-a-têtes.

Our best finiteness result is the following: if there is a finite subset H of B where $G \cup H$ is a SAGBI basis for B then for some step in (3.1), say the jth step, G_j is a SAGBI basis for B. This applies if B has a finite SAGBI basis, simply let H be the finite SAGBI basis for B. The proof of this finiteness result comes later (4.6).

3.6 Remark. There are a few subtleties to finding SAGBI bases which do not apply to finding Gröbner bases. The following is meant to illuminate the situation:

(1) Two familiar and closely related methods, Gröbner$_1$ and Gröbner$_2$, for finding Gröbner bases are sketched.
(2) The SAGBI basis analogs, SAGBI$_1$ and SAGBI$_2$, to Gröbner$_1$ and Gröbner$_2$ are sketched.
(3) SAGBI$_1$ is the SAGBI basis construction presented above (3.1). Thus — at least mathematically speaking — SAGBI$_1$ leads to a SAGBI basis.
(4) An example shows that SAGBI$_2$ does not always lead to a SAGBI basis.

3.7. Gröbner$_1$: Suppose G is a (possibly infinite) subset of $k[\mathbf{X}]$.

Initialize: Set $G_0 = G$.

Inductive step; j^{th} stage with G_j, to $j + 1$ stage with G_{j+1}:
1. Let S_j be the set of G_j s-pairs.
2. For each $(L, R) \in S_j$ let $f(L, R) \in k[\mathbf{X}]$ be a final reductum of $S(L, R)$ over G_j. (The final reductum from any reduction of $S(L, R)$ over G_j will do.) Let F_j be the subset of $k[\mathbf{X}]$ consisting of the $f(L, R)$'s which are not zero.
3. Let $G_{j+1} = F_j \cup G_j$.

Finalize: Set $G_\infty = \bigcup G_j$.

3.8. Gröbner$_2$: Suppose G is a (possibly infinite) subset of $k[\mathbf{X}]$.

Initialize: Set $G_0 = G$.

Inductive step; j^{th} stage with G_j, to $j + 1$ stage with G_{j+1}:
1. Let S_j be set of G_j S-pairs.
2. For each $(L, R) \in S_j$ let $f(L, R) \in k[\mathbf{X}]$ be a final reductum of $S(L, R)$ over G_j. (The final reductum from any reduction of $S(L, R)$ over G_j will do.) If all $f(L, R) = 0$, let F_j be the empty set; otherwise, let $F_j = \{f(L, R)\}$ for one *non-zero* $f(L, R)$.
3. Let $G_{j+1} = F_j \cup G_j$.

Finalize: Set $G_\infty = \bigcup G_j$.

When G is a finite set Gröbner$_1$ always reaches a j where $G_j = G_{j+1}$. This G_j also equals G_∞. Thus by keeping track of G_j from one step to the next, Gröbner$_1$ becomes an algorithm *when G is finite*. The same applies to Gröbner$_2$. Gröbner$_1$ and Gröbner$_2$ are (pessimized[18] versions of) familiar techniques to compute Gröbner bases. The SAGBI basis construction presented above (3.1) — call it SAGBI$_1$ — is the SAGBI analog to Gröbner$_1$. By (3.5), SAGBI$_1$ leads to a SAGBI basis for G. Here is SAGBI$_2$, the SAGBI analog to Gröbner$_2$.

3.9. SAGBI$_2$: Suppose G is a (possibly infinite) subset of $k[\mathbf{X}]$.

Initialize: Set $G_0 = G$.

Inductive step; j^{th} stage with G_j, to $j + 1$ stage with G_{j+1}:
  1. Let T_j be a set which generates the G_j tête-a-têtes.
  2. For each $(L, R) \in T_j$ let $f(L, R) \in k[\mathbf{X}]$ be a final subductum of $T(L, R)$ over G_j. (The final subductum from any subduction of $T(L, R)$ over G_j will do.) If all $f(L, R) \in k$ let F_j be the empty set; otherwise, let $F_j = \{f(L, R)\}$ for one $f(L, R)$ not lying in k.
  3. Let $G_{j+1} = F_j \cup G_j$.

Finalize: Set $G_\infty = \bigcup G_j$.

The following is an example where SAGBI$_2$ does not lead to a SAGBI basis for G. The example concludes the discussion of the *subtlety* first mentioned at (3.6).

3.10 Example. Let B be the subalgebra of $k[X, Y]$ generated by $X \equiv U$, $Y^2 + XY \equiv V$ and $X^3Y \equiv W_2$. Since B is generated by homogeneous elements, B is a graded subalgebra of $k[X, Y]$. Let B have a term ordering with $Y > X$ and let $G_2 = \{U, V, W_2\}$. Renumbering SAGBI$_2$ to start with G_2 instead of G_0 eliminates an offset by 2 in what follows.

LT $U^6V = $ LT $W_2^2 = X^6Y^2$ and hence is a tête-a-tête. (More correctly, the pair of exponent functions (L, R) on B is a tête-a-tête. L and R are the exponent functions which take the value zero except for: $L(U) = 6$, $L(V) = 1$, $R(W_2) = 2$.)

Include this tête-a-tête among the generators for the G_2 tête-a-têtes.

$T(U^6V, W_2^2) = U^6V - W_2^2 = X^7Y \equiv W_3$. Notice what W_3 does not subduct over G_2. According to SAGBI$_2$ set, $G_3 = \{U, V, W_2, W_3\}$. Let $W_i = X^{2^i-1}Y$ and suppose by induction that $G_t = \{U, V, W_2, W_3, \cdots, W_t\}$. LT $U^{2^{t+1}-2}V = $ LT $W_t^2 = X^{2^{t+1}-2}Y^2$ and hence is a tête-a-tête. Include this tête-a-tête among the generators for the G_t tête-a-têtes. $T(u^{2^{t+1}-2}V, W_t^2) = U^{2^{t+1}-2}V - W_t^2 = X^{2^{t+1}-1}X \equiv W_{t+1}$. Notice that W_{t+1} does not subduct over G_t. According to SAGBI$_2$, set $G_{t+1} = \{U, V, W_2, W_3, \cdots, W_t, W_{t+1}\}$.

This completes the induction step and shows that SAGBI$_2$ yields: $G_2, G_3, G_4, \cdots$. Thus SAGBI$_2$ does not terminate after a finite number of steps. On the other hand, $X = $ LT X and $Y^2 = $ LT V. By (4.7) and (4.9) (the key point being that $k[X, Y]$ is integral over gr B) it follows that B has a finite SAGBI basis. By (4.6) it follows that (3.1), SAGBI$_1$, terminates after a finite number of steps.

---

[18]Meaning: *unoptimized.*

4. The big picture/grade school

In Buchberger theory one relates ideals in $k[\mathbf{X}]$ to homogeneous ideals in $k[\mathbf{X}]$ by passing from the not necessarily homogeneous ideal to the homogeneous ideas generated by the lead terms of the original ideal. This is easily SAGBI-ized with similar useful results. As usual we assume $k[\mathbf{X}]$ has a term ordering.

4.1 Definition. Suppose B is a subalgebra of $k[\mathbf{X}]$. $\operatorname{gr} B$ denotes the homogeneous subalgebra of $k[\mathbf{X}]$ spanned over k by the lead terms of all elements of B.

4.2 Remark. A mathematical treatment (and extension) of the notion of term orderings is the *theory of graded structures* presented in [12]. This deals with filtered algebras and their associated graded algebras. The theory developed there applies to $k[\mathbf{X}]$ and other rings. For $k[\mathbf{X}]$, term orderings give rise to filtrations, but there are additional filtrations which do not arise from term orderings. A filtered algebra A has an associated graded algebra $\operatorname{gr} A$. Ideals and subalgebras of A have induced filtrations which give rise to homogeneous or graded ideals and subalgebras of $\operatorname{gr} A$. Suppose $A = k[\mathbf{X}]$ has a filtration which arises from a term ordering. Let I be an ideal of A and let B be a subalgebra of A. The passage from A, I and B to $\operatorname{gr} A$, $\operatorname{gr} I$ and $\operatorname{gr} B$ may be achieved within the theory of graded structures or by the conventional manner of passing to the ideal or subalgebra spanned by the lead terms of I or B. The resulting graded objects are equivalent. Since we have used term orderings rather than the theory of graded structures up to this point, we use lead terms to form gr rather than the theory of graded structures. Further remarks and an example concerning graded structures appear at the end of this section.

The notion of monoid algebra can be found in standard algebra books such as [8, p. 179]. $\operatorname{gr} B$ is a monoid algebra on the monoid of lead terms of B. This simply comes down to the fact that sets of terms are linearly independent in the polynomial ring. If M is a multiplicative commutative monoid and I is a subset of M, I is an *ideal* in M if $M * I \subset I$. Some ideal theory for commutative monoids is sketched in [13], including the notion of *Noetherian monoid*.

In Buchberger theory, a set G, which generates an ideal I, is a Gröber basis for I if and only if $\operatorname{gr} G$ generates the homogeneous ideal $\operatorname{gr} I$. The obvious SAGBI analog holds:

4.3 Proposition. *Let $S \subset k[\mathbf{X}]$ generate the subalgebra B. S is a SAGBI basis for B if and only if the lead terms of S generate $\operatorname{gr} B$.*

PROOF: By the definition of SAGBI basis (1.11), a subset $S \subset B$ is a SAGBI basis for B if and only if the lead terms of S generate the monoid of lead terms of B. Since $\operatorname{gr} B$ is a monoid algebra, a subset of the monoid generates the monoid if and only if it generates the monoid algebra.

The next result tells about finite SAGBI bases.

4.4 Proposition. *Let B be a subalgebra of $k[\mathbf{X}]$. The following statements are equivalent:*

a. $\operatorname{gr} B$ *is Noetherian.*

b. $\operatorname{gr} B$ *is a finitely generated algebra (over k).*

c. *The multiplicative monoid of lead terms of B is finitely generated.*

d. B *has a finite SAGBI basis.*

e. *Every SAGBI basis of B has a finite subset which is also a SAGBI basis for B.*

PROOF: $\operatorname{gr} B$ is a graded subalgebra of $k[\mathbf{X}]$ and so the zeroth component of $\operatorname{gr} B$ is a field k. It follows from a standard characterization of when graded rings are Noetherian [8, p. 239, 7.2], that $\operatorname{gr} B$ is Noetherian if and only if $\operatorname{gr} B$ is finitely generated over k. Since $\operatorname{gr} B$ is a monoid algebra, it is finitely generated over k if and only if the monoid is finitely generated. Thus parts (a), (b) and (c) are equivalent. By the definition of SAGBI basis (1.11), a subset $S \subset B$ is a SAGBI basis for B if and only if the lead terms of S generate the monoid of lead terms of B. Thus parts (c) and (d) are equivalent. The equivalence of (d) and (e) stems from the first part of the lemma which follows. ∎

This lemma is the key to the finiteness of SAGBI basis construction promised earlier.

4.5 Lemma. *Let M be a commutative monoid. The following statements are equivalent.*
a. *M is finitely generated.*
b. *If $S \subset M$ and S generates M then there is finite set $F \subset S$ where F generates M.*
c. *If $\{M_\alpha\}$ is a directed[19] set of submonoids of M where $\bigcup_\alpha M_\alpha = M$ then for some β, $M_\beta = M$.*

Let $G \subset M$, the following statements are equivalent:
d. *There is a finite set $T \subset M$ where $G \cup T$ generates M.*
e. *If $S \subset M$ and $G \cup S$ generates M then there is finite set $F \subset S$ where $G \cup F$ generates M.*
f. *If $\{M_\alpha\}$ is a directed set of submonoids of M where $\bigcup_\alpha M_\alpha = M$ and $\bigcap_\alpha M_\alpha \supset G$ then for some β, $M_\beta = M$.*

PROOF: Letting G be the empty set shows that it is only necessary to prove the equivalence of (d), (e) and (f).

(f) implies (e). Suppose $G \cup S$ generates M. For each finite subset $F_\alpha \subset S$, let M_α be the submonoid of M generated by $G \cup F_\alpha$. Clearly $\{M_\alpha\}$ satisfies (f). Thus there is a β where $M_\beta = M$ and for the finite subset F_β of S, $G \cup F_\beta$ generates M.

(d) implies (f). Let T be any finite set where $G \cup T$ generates M. Since $\bigcup_\alpha M_\alpha = M$ in (f), each element of T lies in an M_α. By the finiteness of G and the *directed* condition on $\{M_\alpha\}$ it follows that there is a single β where $T \subset M_\beta$. By hypothesis (f), G also lies in M_β. Since $G \cup T$ generates M and M_β is a submonoid it follows that $M_\beta = M$.

(e) implies (d) is clear. ∎

The following theorem shows that the SAGBI basis construction method (3.1) is *as good as it gets*. I. e. if G is the initial set and there is any finite set H where $G \cup H$ is a SAGBI basis then (3.1) terminates in a finite number of steps.

4.6 Proposition. *Suppose G is a (possibly infinite) subset of $k[\mathbf{X}]$ and B is the subalgebra of $k[\mathbf{X}]$ generated by G. We use the notation of (3.1). If there is a finite subset H of B where $G \cup H$ is a SAGBI basis for B then for some step in (3.1), say the jth step, G_j is a SAGBI basis for B. This applies if B has a finite SAGBI basis, simply let H be the finite SAGBI basis for B.*

PROOF: G_∞ is a SAGBI basis for B by (3.5). Let $S = G_\infty \setminus G$. By the previous lemma there is a finite subset $F \subset S$ where the lead terms of $G \cup F$ generate the monoid of lead

---

[19] By directed we mean that given M_α and M_β there is an M_γ with: $M_\alpha \subset M_\gamma \supset M_\beta$.

terms of B. Since F is finite there is a j where $F \subset G_j$. Since $G_j \supset G$, it follows that G_j is a SAGBI basis. ∎

The next result gives a sufficient condition for B to have a finite SAGBI basis.

4.7 Proposition. *Suppose B is a subalgebra of $k[\mathbf{X}]$ and C is a finitely generated subalgebra of $k[\mathbf{X}]$ containing $\operatorname{gr} B$. If C is integral over $\operatorname{gr} B$ then B has a finite SAGBI basis. In particular, if $k[\mathbf{X}]$ is integral over $\operatorname{gr} B$ then B has a finite SAGBI basis.*

PROOF: By [1, p. 81, 7.8], it follows that $\operatorname{gr} B$ is finitely generated over k. Hence, (4.4) implies that B has a finite SAGBI basis. ∎

4.8 Corollary. *Every subalgebra of a polynomial ring in one variable (over a field) has a finite SAGBI basis.*

PROOF: Let $k[\mathbf{X}]$ denote the polynomial ring in one variable. The subalgebra k has the empty set as a finite SAGBI basis (1.12). If B is a subalgebra of $k[X]$ other than k then $\operatorname{gr} B$ contains a monic non-constant polynomial $f(X)$. Hence, X satisfies the monic polynomial $f(Z) - f(X) \in B[Z]$ and $k[X]$ is integral over $\operatorname{gr} B$. ∎

(4.7) raises the issue of $k[\mathbf{X}]$ being integral over $\operatorname{gr} B$. This is characterized as follows:

4.9. INTEGRAL OVER $\operatorname{gr} B$: $k[X_1, \cdots, X_n]$ is integral over $\operatorname{gr} B$ if and only if for each $i = 1, \cdots, n$ there is a positive integer d_i and an element $b_i \in B$ whose lead term is $X_i^{d_i}$.

PROOF: If such d_i's and b_i's exist then $X_i^{d_i} = \operatorname{LT} b_i \in \operatorname{gr} B$ and $k[X_1, \cdots, X_n]$ is integral over $\operatorname{gr} B$. Conversely, say $k[X_1, \cdots, X_n]$ is integral over $\operatorname{gr} B$. Then X_1 satisfies an integral equation of the form:

4.10 $$X_1^m + g_1 X_1^{m-1} + \cdots + g_{m-1} X_1 + g_m = 0 \quad \text{with} \quad \{g_i\} \subset \operatorname{gr} B$$

Each g_i is a sum of monomials. At least one g_i must have a non-zero monomial with term of the form $X_1^{d_1}$, otherwise nothing would cancel the X_1^m at (4.10). Since $\operatorname{gr} B$ is a monoid algebra, each of the individual terms of the g_i's lie in $\operatorname{gr} B$. Thus, $\operatorname{gr} B$ contains $X_1^{d_1}$ and B has an element whose lead term is $X_1^{d_1}$. Similarly for the other X_i's. ∎

If B is a graded algebra over the field k and B_i denotes the ith graded (homogeneous) component of B, the Hilbert function of B is defined as the function $H_B : \mathbf{N} \to \mathbf{N}$ with $H_B(i) = \dim_k B_i$. Of course, this is simply another way to look at the infinite sequence of non-negative integers: $(\dim_k B_0, \dim_k B_1, \dim_k B_2, \cdots)$. Still another way is given by the "generating function" in the power series ring $\mathbf{Z}[[t]]$: $(\dim_k B_0) + (\dim_k B_1)t + (\dim_k B_2)t^2 + \cdots$. $\mathbf{Z}$ denotes the integers. This power series is called the Hilbert-Poincaré series of B or simply the Poincaré series of B.[20] In the next example we shall use standard results and techniques concerning the Hilbert-Poincaré series. One, easily verified, result we shall use is that if C is a homogeneous subalgebra of $k[\mathbf{X}]$ then the Hilbert-Poincaré series of C is the same as the Hilbert-Poincaré series of $\operatorname{gr} C$.

(1.20) is an example of a subalgebra B which does not have a finite SAGBI basis with any term ordering. As pointed out just below (1.20), $k[X,Y]$ is integral over B. Since B does not have a finite SAGBI basis, it follows from (4.7) that $k[X,Y]$ is not integral

---

[20] The Hilbert function/Poincaré series is one of three fundamental approaches to dimension theory. The other two approaches are: the (prime) ideal theoretic (Krull dimension) approach and the transcendence degree approach. The three approaches are described and related in [1, chapter 11].

over gr B. Thus (1.20) is also an example where $k[X, Y]$ is integral over B but $k[X, Y]$ is not integral over gr B, for any term ordering. The next example presents a finitely generated graded subalgebra of $k[X, Y]$ and two term orderings. The subalgebra has a finite SAGBI basis with the first term ordering but does not have a finite SAGBI basis with the second term ordering.

4.11 Example of a finitely generated graded subalgebra which does/does not have a finite SAGBI basis. Let B be the subalgebra of $k[X, Y]$ (finitely) generated by $X, XY - Y^2, XY^2$. This example goes beyond presenting a finitely generated subalgebra with the stated properties. Where (1.20) used direct computation, this example illustrates the Hilbert-Poincaré series and other techniques applied to SAGBI bases.

To begin, let us determine the Hilbert function of B. By ordinary Gröbner basis methods which use tag variables to ascertain the relations among polynomials, [15], [5, section 3], it is easily determined that if: $u = X$, $v = XY - Y^2$, and $w = XY^2$ then u, v, w satisfy the minimal relation: $u^2v^2 - u^3w + 2uvw + w^2 = 0$. In other words given the map:

$$\pi \colon k[U, V, W] \longrightarrow k[X, Y]$$

determined by:

$$U \longrightarrow X, \quad V \longrightarrow XY - Y^2, \quad W \longrightarrow XY^2$$

the kernel is the ideal generated by: $U^2V^2 - U^3W + 2UVW + W^2$. Let $k[U, V, W]$ have the unique grading where U has degree 1, V has degree 2 and W has degree 3. Then π is a homogeneous map. $k[U, V, W]$ is the tensor product of $k[U]$, $k[V]$, $k[W]$ with Hilbert-Poincaré series $1/(1 - t)$, $1/(1 - t^2)$, $1/(1 - t^3)$, respectively.[21] Thus $k[U, V, W]$ has Hilbert-Poincaré series $1/(1 - t)(1 - t^2)(1 - t^3)$. By usual Hilbert-Poincaré series manipulations, the fact that we are factoring out an ideal generated by a homogeneous element of degree 6 from $k[U, V, W]$, implies that the quotient has Hilbert-Poincaré series $(1 - t^6)/(1 - t)(1 - t^2)(1 - t^3)$. Now a little manipulation:

$$
\begin{aligned}
4.12 \qquad (1 - t^6)/(1 - t)(1 - t^2)(t - t^3) &= (1 + t^3)/(1 - t)(1 - t^2) \\
&= (1 - t + t^2)/(1 - t)^2 \\
&= (1 - t + t^2)(1 + 2t + 3t^2 + 4t^3 + \cdots) \\
&= 1 + t + 2t^2 + 3t^3 + \cdots + nt^4 + \cdots
\end{aligned}
$$

Since π maps $k[U, V, W]$ onto B, (4.12) is the Hilbert function of B.

Let $k[X, Y]$ have a term ordering where $Y > X$. Then $XY - Y^2$ has lead term Y^2 and both X and Y^2 lie in gr B. Hence, $k[X, Y]$ is integral over gr B. By (4.7), it follows that B has a finite SAGBI basis. Thus by (4.6), (3.1) will terminate after a finite number of steps with a SAGBI basis. The Hilbert-Poincaré series enables us to easily prove that a good guess, $\{X, XY - Y^2, X^2Y\}$, is indeed a SAGBI basis. Note, $X^2Y = X(XY - Y^2) - XY^2 \in B$.

It is easily seen that gr $B \supset k[X, X^2, X^2Y]$. The minimal relation among: $a = X$, $b = Y^2$ and $c = X^2Y$ is: $a^4b = c^2$. The same calculation used for computing the Hilbert function for B (and hence gr B) shows that (4.12) is the Hilbert-Poincaré series

---

[21] The exponents 2 and 3 for t arise because U and V have degree 2 and 3.

of $k[X, Y^2, X^2Y]$. Since $\operatorname{gr} B \supset k[X, Y^2, X^2Y]$ the fact that they have the same Hilbert-Poincaré series implies that $\operatorname{gr} B = k[X, Y^2, X^2Y]$. Hence, $\{X, XY - Y^2, X^2Y\}$ is a SAGBI basis for B.

Next let $k[X, Y]$ have a term ordering with $X > Y$. For even and odd non-negative integers let:

$$f_{\text{even}} = XY^{\text{even}} \qquad f_{\text{odd}} = ((\text{odd} + 1)/2)XY^{\text{odd}} - Y^{\text{odd}+1}$$

The first three values are: $f_0 = X$, $f_1 = XY - Y^2$ and $f_2 = XY^2$. These elements are the given generators for B. Notice that:

$$f_{\text{even}+1} = (\text{even}/2)f_0 f_{\text{even}} - f_1 f_{\text{even}-1}$$
$$f_{\text{odd}+1} = (2f_0 f_{\text{odd}} - (\text{odd} + 1)f_1 f_{\text{odd}-1})/(\text{odd} - 1)$$

Thus all the f_{even}'s and f_{odd}'s lie in B. Since $X > Y$, the lead term of f_{odd} is XY^{odd}. It follows that $\{X, XY, XY^2, XY^3, \cdots\} \in \operatorname{gr} B$ and hence $k[X, XY, XY^2, XY^3, \cdots] \subset \operatorname{gr} B$. Notice that in degree n, $k[X, XY, XY^2, XY^3, \cdots]$ has k basis consisting of the terms: $X^n, X^{n-1}Y, X^{n-2}Y^2, \cdots, XY^n$. Thus $k[X, XY, XY^2, XY^3, \cdots]$ has Hilbert-Poincaré series given by (4.12). Since $k[X, XY, XY^2, XY^3, \cdots] \subset \operatorname{gr} B$ and they have the same Hilbert-Poincaré series, they are equal. Thus $\operatorname{gr} B$ is not finitely generated and B does not have a finite SAGBI basis. ∎

In example (1.20), a key step of the proof that B does not have a finite SAGBI basis is the direct computational verification that B and hence $\operatorname{gr} B$ does not contain Y^n for any n. The use of the Hilbert-Poincaré series in the previous example obviates a direct computational verification that $\operatorname{gr} B$ does not contain Y^n for any n.

SAGBI bases generalize from term orderings to graded structures in analogy to the generalization of Gröbner bases from term orderings to graded structures in [12]. The next example shows that graded structures *see more* than term orderings. Example (1.20) was shown to have no finite SAGBI basis in the term ordering based SAGBI theory. The next example shows that the graded structure based SAGBI theory *is able to see* a finite SAGBI basis for example (1.20). Rather than go into full detail, this brief example is written for those who already know about graded structures. It is hoped that others will find elements of the example concerning the automorphism α interesting. If not, it may be ignored without jeopardizing a general reading of this paper.

4.13 Graded structure and automorphism example. As in example (1.20), let B be the subalgebra of $k[X, Y]$ generated by $X + Y, XY, XY^2$. Let α be the automorphism of $k[X, Y]$ which sends X to $X + Y$ and Y to $-Y$. Notice that α is its own inverse and α maps $X + Y$ to X and XY to $-XY - Y^2$. Thus $X, XY + Y^2 \in \alpha(B)$. If $k[X, Y]$ has a term ordering with $Y > X$ it follows that $X, Y^2 \in \operatorname{gr} \alpha(B)$. Thus $k[X, Y]$ is integral over $\operatorname{gr} \alpha(B)$ and by (4.7), $\alpha(B)$ has a finite SAGBI basis. This shows that the property of having a finite SAGBI basis is not invariant under automorphism. A term ordering on $k[X, Y]$ gives a graded structure filtration on $k[X, Y]$. The image of this filtration under α^{-1}, which is just α itself, is a graded structure filtration on $k[X, Y]$. The image under α^{-1} of $\alpha(B)$ is of course B. The image under α^{-1} of the finite SAGBI basis for $\alpha(B)$ is a finite SAGBI basis *in the graded structure SAGBI theory* for B. Thus B has no finite SAGBI basis in the term ordering based SAGBI theory but has a finite SAGBI basis in the graded structure based SAGBI theory.

The previous example raises interesting questions for those interested in pursuing the graded structure based SAGBI theory. The most obvious question is: is there a finitely generated subalgebra of $k[\mathbf{X}]$ which has no finite SAGBI basis in the graded structure based SAGBI theory?

5. Homogeneous subalgebras and partial SAGBI bases

In SAGBI theory, just as in Buchberger theory, refinements are possible when dealing with homogeneous elements. This brief section sketches one such refinement. The upshot is that homogeneous SAGBI subalgebra membership determination is algorithmic.

$k[\mathbf{X}]$ is assumed to have a term ordering. We are interested in algorithmically solving:

5.1 Problem. d is a positive integer. G is a finite set of homogeneous polynomials of degree d or less, B is the subalgebra generated by G and f is a polynomial in $k[\mathbf{X}]$ of degree d. *Is f in B?* The SAGBI solution appears at (5.5).

Notice that if E is an exponent function on B with support consisting of homogeneous polynomials then R^E is homogeneous. Thus if (E, F) is a tête-a-tête on B where E and F have homogeneous support then $T(E, F)$ is homogeneous. Consequently, if G consists of homogeneous polynomials and one performs (3.1), then all the polynomials in the G_i's are homogeneous. This establishes:

5.2 Proposition. *If G consists of homogeneous polynomials then (3.1) produces a homogeneous SAGBI basis.* ∎

The solution to (5.1) rests on an algorithm which constructs the low degree elements in a homogeneous SAGBI basis.

5.3. ALGORITHM FOR PARTIAL SAGBI BASIS: Suppose G is a finite subset of $k[\mathbf{X}]$ consisting of homogeneous polynomials of degree d or less.

Initialize: Set $H_0 = G$.

Inductive step; j^{th} stage with H_j, to $j + 1$ stage with $Hj + 1$:
  1. Let U_j be a finite set which generates the H_j tête-a-têtes.
  2. For each $(L, R) \in U_j$ let $f(L, R) \in k[\mathbf{X}]$ be a final subductum of $T(L, R)$ over H_j. (The final subductum from any subduction of $T(L, R)$ over H_j will do.) Let E_j be the subset of $k[\mathbf{X}]$ consisting of the $f(L, R)$'s which do not lie in k and have degree d or less.
  3. Let $H_{j+1} = E_j \cup H_i$.

Finalize: Set $H_\infty = \bigcup H_j$.

5.4 Partial SAGBI basis theorem.

5.4.a. Algorithmicity: *(5.3) stabilizes/terminates after a finite number of steps (with a finite H_∞.)*

5.4.b. Completeness: *The subalgebra generated by G has a SAGBI basis G_∞ for which:*

$$H_\infty = \{g \in G_\infty | \text{ where } \deg g \leq d\}$$

PROOF: (5.4.a) If f and s are non-constant homogeneous elements with the same lead term then for suitable $\lambda \in k : f - \lambda s$ is a subduction of f over $\{s\}$. Thus the lead terms of elements in each E_j must differ from the lead terms of elements in H_j. Since there are only a finite number of distinct lead terms of degree d or less there is a point, say D, beyond which E_j is empty for $j > D$. Thus $H_\infty = G \cup E_0 \cup E_1 \cup \cdots E_D$.

(5.4.b) Starting with G perform (3.1) with the following refinement: at stage (2) in the inductive step let F_j' consist of the final subductums of degree d or less which do not lie in k and let F_j'' consist of the final subductums of degree $d + 1$ or greater. Thus $F_j = F_j' \cup F_j''$ and the SAGBI basis G_∞ given by (3.1) decomposes:

$$G_\infty = (G \cup F_j') \cup F_j''.$$

The elements of G_j of degree higher than d cannot lead to elements of F_j'. Thus the E_j's of (5.3) are the same as the F_j''s and $H_\infty = (G \cup F_j')$. ∎

5.5. SOLUTION OF (5.1): Starting with G perform (5.3) to get H_∞. f lies in B if and only if f subduces to an element of k over H_∞. This follows from (1.16.a) and (5.4.b) together with the fact that degree considerations imply that if f subduces over G_∞ then all the subduction must have been over H_∞.

References

1. ATIYAH, M. F., MACDONALD, I. G., "Introduction to Commutative Algebra," Addison-Wesley, 1969.
2. FAGES, F., *Associative-commutative unification*, J. Symb. Comp. **3** (1987), 257–275.
3. CLAUSEN, M., FORTENBACHER, A., *Efficient solution of linear diophantine equations*, Interner Bericht 32/87, Universität Karlsruhe (1987).
4. CLAUSEN, M., FORTENBACHER, A., *Efficient solution of linear diophantine equations*, J. Symb. Comp. **8** (1989), 201–216.
5. GIANNI, P., TRAGER, B., ZACHARIAS, G., *Gröbner bases and primary decomposition of polynomial ideals*, J. Symb. Comp **6** (1988), 149–167.
6. KAPUR, D., MADLENER, K., *A completion procedure for computing a canonical basis for a k-subalgebra*, in "Computers and Mathematics (Cambridge MA 1989)," Springer, New York, 1989, pp. 1–11.
7. LAMBERT, J., *Une borne pour les generateurs des solutions entieres positives d'une equation diophantine lineare.*, University de Paris-Sud, Laboratoire de Recherche en Informatique, Rap. 334, Orsay (1987).
8. LANG, S., "Algebra," Addison-Wesley, Reading, Mass., 1984.
9. MORA, T., *Gröbner bases for non-commutative polynomial rings, AECC 3*, in "Lecture Notes in Computer Science 229," Springer, 1986, pp. 353–362.
10. LANKFORD, D., *New non-negative integer basis algorithms for linear equations with integer coefficients.*, Unpublished manuscript.
11. ROBBIANO, L., *Term orderings on the polynomial ring. Proc. EUROCAL 85, II*, in "Lecture Notes in Computer Science 204," Springer, 1985, pp. 513–517.
12. ROBBIANO, L., *On the theory of graded structures*, J. Symb. Comp. **2** (1986), 139–170.
13. ROBBIANO, L., *Introduction to the theory of Gröbner bases*, Queen's Papers in Pure and Applied Mathematics, V 5, No. 80 (1988).

14. ROBBIANO, L., *Computing a SAGBI basis of a k-subalgebra*, Presented at Mathematical Sciences Institut Workshop on Gröbner bases, Cornell University (1988).

15. SPEAR, D., *A constructive approach to commutative ring theory*, in "Proceedings 1977 MACSYMA User's Conference," 1977, pp. 369–376.

16. SHANNON, D., SWEEDLER, M., *Using Gröbner bases to determine algebra membership, split surjective algebra homomorphisms and determine birational equivalence*, J. Symb. Comp. **6** (1988), 267–273.

17. SWEEDLER, M., *Ideal bases and valuation rings*, Preprint (1986).

18. SWEEDLER, M., *Ideal bases and valuation rings: an overview*, Preprint (1987).

19. ZHANG, H., *Speeding up basis generation of homogeneous linear deophantine equations*, Unpublished manuscript (1989).

Dipartimento di Matematica, Universita di Genova, Via L. B. Alberti 4, 16132 Genova, Italy

320 White Hall, Department of Mathematics, Cornell University, Ithaca NY 14853, U.S.A.

Flatness and Ideal-Transforms of Finite Type

Dedicated to Professor David Rees on his seventieth birthday

1. Introduction

Ideal-transforms were introduced by Nagata in [10] and [11] and they proved to be very useful in his series of papers related to the Fourteenth Problem of Hilbert. On the other side, their finiteness resp. Noetherianness is related to several problems in commutative algebra, see [14] for one of them. The starting point of our investigations here is the following observation done by Richman in [12]. Let A denote a Noetherian domain with $Q(A)$ its quotient field. An intermediate ring $A \subseteq B \subseteq Q(A)$ flat over A is a Noetherian ring. Therefore, a question related to the finiteness of the ideal-transform

$$T_I(A) = \{r \in Q(A) : I^n r \subseteq A \text{ for some } n \in \mathbb{N}\},$$

$I \neq (0)$ an ideal of A, is to determine when $T_I(A)$ is an A-flat module.

Theorem (cf. (3.4) and (2.3)). *For a regular ideal I of a commutative Noetherian ring A the following conditions are equivalent:*
i) $T_I(A)$ *is flat over A.*
ii) $\operatorname{cd}_A J = 1$, *where $J = xA : (xA : \langle I \rangle)$ for $x \in I$ a nonzero divisor.*
iii) $\operatorname{Spec} A \setminus V(J)$, *$J$ as before, is an affine scheme.*

Here $\operatorname{cd}_A J$, J an ideal of A, denotes the cohomological dimension of A with respect to J, see Section 2 for the definition. Note that $\operatorname{cd}_A J$ was introduced by Hartshorne in [6]. In Section 2 there is a characterization when $\operatorname{cd}_A J \leq 1$. In general $\operatorname{ht} J \leq \operatorname{cd}_A J \leq \dim A$ and it is rather hard to describe the precise value of $\operatorname{cd}_A J$, see [6] for some results in this direction.

Corollary (cf. (3.5)). *Let I and A be as above. Suppose $T_I(A)$ is flat over A. Then $T_I(A)$ is an A-algebra of finite type.*

Thus a flat ideal-transform is not merely a Noetherian ring but also an algebra of finite type. The converse of the Corollary does not hold as shown by an example in Section 2.

The Hartshorne-Lichtenbaum vanishing theorem, see [6], provides when $\operatorname{cd}_A I < \dim A$ for an ideal I of A. Thus, in the case of an ideal I of height one in a two-dimensional local ring $(A, \mathfrak{m})$ this leads to an intrinsic characterization when $\operatorname{cd}_A I = 1$. Pursuing further the point of view of two-dimensional local rings it follows:

Theorem (cf. (4.1)). *For a regular ideal I of height one in a two-dimensional local ring $(A, \mathfrak{m})$ the following conditions are equivalent:*
i) $T_I(A)$ *is a flat A-module.*
ii) $\dim \hat{A}/(I\hat{A} + \mathfrak{p}) = 1$ *for all two-dimensional prime ideals $\mathfrak{p} \in \operatorname{Ass} \hat{A}$, $\hat{A}$ denotes the completion of A.*

iii) $T_I(A)$ *is an A-algebra of finite type.*

In connection with Zariski's generalization of Hilbert's Fourteenth Problem it yields a complete picture when ideal-transforms of two-dimensional local rings are of finite type. This is also related to recent research in [4].

In (4.4) we conclude with examples when $T_I(A)$ is finitely resp. not finitely generated as an A-algebra. In the terminology we follow [9].

2. Cohomological dimension one

Let I denote an ideal of a commutative ring A. For our purposes we need the local cohomology functors $H_I^i(\ ), i \in \mathbb{Z}$, of A with respect to I. See Grothendieck [5] for basic facts about them. In the following let

$$\mathrm{cd}_A\, I = \sup\{i \in \mathbb{Z} : H_I^i(F) \neq 0, F \text{ an } A\text{-module}\},$$

the cohomological dimension of I. For the importance of this notion see Hartshorne's article [6]. For a Noetherian ring A

$$\mathrm{cd}_A\, I = \sup\{i \in \mathbb{Z} : H_I^i(A) \neq 0\}$$

and $\mathrm{ht}\, I \leq \mathrm{cd}_A\, I \leq \dim A$. In general it seems to be hopeless to characterize $\mathrm{cd}_A\, I$ in more intrinsic terms of I and A.

Related to the local cohomology let us consider the covariant, additive, A-linear, left exact functor

$$T_I(\ ) = \varinjlim_n \mathrm{Hom}_A(I,\ ),$$

where the maps in the direct systems are induced by inclusions. For the right derived functors $R^iT_I(\ ), i \in \mathbb{Z}$, of $T_I(\ )$ there are an exact sequence

$$0 \longrightarrow H_I^0(F) \longrightarrow F \longrightarrow T_I(F) \longrightarrow H_I^1(F) \longrightarrow 0$$

and isomorphisms $R^iT_I(F) \simeq H_I^{i+1}(F), i \geq 1$, where F is an A-module. We call $T_I(F)$ the ideal-transform of F with respect to I. Suppose I is a regular ideal. Then

$$T_I(A) = \bigcup_{n \geq 1} I^{-n}, \quad I^{-n} = \{r \in Q(A) : I^n r \subseteq A\},$$

where $Q(A)$ denotes the full ring of quotients of A. In this situation $T_I(A)$ is the classical definition of the ideal-transform. $T_I(A)$ is the ring of global sections over $\mathrm{Spec}\, A \setminus V(I)$.

In the following we give a characterization when $\mathrm{cd}_A\, I \leq 1$ in terms of $T_I(\ )$ and the local cohomolgy modules.

(2.1) Proposition. *For an ideal I of A the following conditions are equivalent:*
i) $\mathrm{cd}_A\, I \leq 1$.
ii) $H_I^2(F) = 0$ *for all A-modules F.*
iii) $T_I(\ )$ *is an exact functor.*

iv) *The canonical homomorphism* $T_I(A) \otimes_A F \longrightarrow T_I(F)$ *is an isomorphism for all A-modules F.*

v) $T_I(A) = IT_I(A)$.

PROOF: First note that the implications i) $\implies$ ii) $\implies$ iii) $\implies$ iv) are easy to show. In order to prove iv) $\implies$ v) note that

$$T_I(A)/I\, T_I(A) \simeq T_I(A) \otimes_A A/I \simeq T_I(A/I) = 0$$

because $\mathrm{Supp}_A A/I = V(I)$. To finish the proof let us show v) $\implies$ i). Let $i \geq 2$. Then

$$H_I^i(A) \simeq H_I^i(T), \quad T := T_I(A),$$

as follows by the above exact sequence and $H_I^i(H_I^j(A)) = 0$ for $i > 0$ and $j = 0, 1$. By virtue of [5], Corollary 5.7, we see that $H_I^i(T) \simeq H_{IT}^i(T) \simeq H_T^i(T) = 0$. That is, $\mathrm{cd}_A I \leq 1$, as required.

The condition v) of (2.1) has do to with Hilbert's Fourteenth Problem, see [10] and [11], as follows by view of the next result.

(2.2) Corollary. *Suppose I is a regular ideal of A such that* $T_I(A) = I\, T_I(A)$. *Then* $T_I(A)$ *is an A-algebra of finite type.*

PROOF: Because of the assumption there exists a relation

$$1 = \sum_{i=1}^{n} m_i x_i$$

with $m_i \in I, x_i \in T := T_I(A)$ for $i = 1, \ldots, n$. Now we claim $T = A[x_1, \ldots, x_n]$. Let $y \in T$. Choose an integer n such that $I^n y \subseteq A$. The n-th power of the above relation yields

$$1 = \sum_{|a|=n} m_a \underline{x}^a, a = (a_1, \ldots, a_n),$$

with $m_a \in I^n$. Therefore,

$$y = \sum_{|a|=n} (m_a y)\underline{x}^a \in A[x_1, \ldots, x_n],$$

as required.

Of course, the converse of the Corollary above does not hold. Let k be field with w, x, y, z indeterminates over k. Put

$$A = k[w, x, y, z]/(wz - xy) \text{ and}$$
$$\mathfrak{p} = (x, z)A.$$

Note that A is the coordinate ring of a non-singular quadric. The prime ideal $\mathfrak{p}$ with ht $\mathfrak{p} = 1$ corresponds to a line on the quadric. Then

$$(xA : z)x^{-1} = \mathfrak{p}^{-1} = \mathfrak{p}^{-n} \quad \text{for all } n \geq 1,$$

as easily seen. Therefore $T = T_{\mathfrak{p}}(A) \simeq k[x, z, w/x]$ is of finite type over A with $T \neq \mathfrak{p}T$. That is, $\mathrm{cd}_A \mathfrak{p} = 2$. Whence, $\mathrm{cd}_A I = 1$ does not hold in general for an ideal I of height one.

(2.3) Remark. We will add a geometric description of $\mathrm{cd}_A I \leq 1$ for an ideal I of A. To this end recall that $T = I\,T, T := T_I(A)$, holds if and only if $\operatorname{Spec} A \setminus V(I)$ is an affine scheme, see [6].

Let us continue with a permanence behaviour of $\mathrm{cd}_A I \leq 1$ by passing to a ring extension $A \subseteq B$.

(2.4) Proposition. *Suppose $T_I(A) = I\,T_I(A)$ for an ideal I of a commutative Noetherian ring A. Then $T_{IB}(B) = I\,T_{IB}(B)$ for a commutative ring extension $A \subseteq B$. The converse holds provided B is a finitely generated A-module.*

PROOF: By [5], Corollary 5.7, there are isomorphisms

$$H_I^i(F^A) \simeq H_{IB}^i(F)^A, i \in \mathbb{Z},$$

for an B-module F. By (2.1) the assumption yields $H_{IB}^i(F) = 0$ for any B-module F and $i \geq 2$. Now the same argument as given in the proof of (2.1) shows the claim. Under the additional assumption on B Chevalley's Theorem, see [6], yields that $\operatorname{Spec} B \setminus V(IB)$ is affine provided $\operatorname{Spec} A \setminus V(I)$ is affine.

By the altitude, alt I, of an ideal I of A denote the maximum of the heights of minimal prime ideals of I. Furthermore, let A' denote the integral closure of A.

(2.5) Corollary. *Suppose $T_{IA'}(A') = I\,T_{IA'}(A')$ (e.g., $T_I(A) = I\,T_I(A)$) for a non-zero ideal I of a commutative Noetherian domain A. Then $\operatorname{alt} IA' = 1$, i.e., each minimal prime divisor of IA' has height one.*

PROOF: Assume the contrary. Then there is a minimal prime ideal $P \supseteq IA'$ such that $\operatorname{ht} P > 1$. Because A'_P is a Krull domain and $\operatorname{ht} IA'_P > 1$ it follows $T_{IA'_P}(A'_P) = A'_{P'}$ see [10]. On the other hand

$$T_{IA'_P}(A'_P) \simeq T_{IA'}(A') \otimes_{A'} A'_P.$$

By our assumptions it follows $A'_P = IA'_P$, a contradiction to the choice of P.

It will be shown that under more restrictive assumptions on A the converse of (2.5) is also true, see Section 4.

3. Flat ideal-transforms

For an ideal I of A it turns out, see (2.1), that $T_I(A)$ is a flat A-module provided $\mathrm{cd}_A I \leq 1$. In this section we will give a complete picture when $T_I(A)$ is a flat A-module. Put $\operatorname{grade}_A I = \inf\{i \in \mathbb{Z} : H_I^i(A) \neq 0\}$ for a proper ideal I of A. Note that $\operatorname{grade}_A I \geq 2$ if and only if the canonical homomorphism $A \longrightarrow T_I(A)$ is an isomorphism. Let $x \in I$ denote a non-zero divisor and $xA : \langle I \rangle = \bigcup_{n \geq 1} xA : I^n$. Then

$$xA : \langle I \rangle = xT_I(A) \cap A \quad \text{and}$$
$$\operatorname{Ass}_A(xA : \langle I \rangle / xA) = \operatorname{Ass} A / xA \cap V(I) = \operatorname{Ass} I^{-1} / A.$$

Whence these sets of associated prime ideals do not depend on the particular choice of $x \in I$.

(3.1) Proposition. *Let $x \in I$ be a non-zero divisor. Then the following conditions are eqivalent:*
i) *$T_I(A)$ is a flat A-module.*
ii) *$T_I(A) = \mathfrak{p}T_I(A)$ for all $\mathfrak{p} \in \operatorname{Ass} A/xA \cap V(I)$.*

PROOF: First we show i) $\Longrightarrow$ ii). Let $\mathfrak{p} \in \operatorname{Ass} A/xA \cap V(I)$. Suppose that $\mathfrak{p}T, T := T_I(A)$, is a proper ideal. By the Going Down Theorem it follows $\mathfrak{p} = P \cap A \supseteq I$ for a minimal prime ideal $P \supseteq \mathfrak{p}T$. Then there is the following homomorphism

$$A_{\mathfrak{p}} \longrightarrow T_P \subseteq \mathbb{Q}(A_{\mathfrak{p}}).$$

Then $A_{\mathfrak{p}} = T_{\mathfrak{p}} = T_P$ because it is faithfully flat. That is, $\operatorname{grade}_{A_{\mathfrak{p}}} IA_{\mathfrak{p}} \geq 2$ and $\mathfrak{p} \notin \operatorname{Ass} A/xA \cap V(I)$, contracting the choice of $\mathfrak{p}$. Therefore, $\mathfrak{p}T = T$ as required.

In order to prove ii) $\Longrightarrow$ i) it is enough to show that $A_{\mathfrak{p}} \longrightarrow T_P$ is a flat homomorphism for all $P \in \operatorname{Spec} T$ and $\mathfrak{p} = P \cap A$. If $\mathfrak{p} \not\supseteq I$, then $A_{\mathfrak{p}} = T_{\mathfrak{p}} = T_P$ as easily seen. Now let $\mathfrak{p} \supseteq I$. Then $\mathfrak{p}T \subseteq P$, i.e., $\mathfrak{p}T$ is a proper ideal. By the assumption $\mathfrak{p} \notin \operatorname{Ass} A/xA$ and $A_{\mathfrak{p}} = T_{\mathfrak{p}} = T_P$. That is, T_P is flat over $A_{\mathfrak{p}}$ for all $P \in \operatorname{Spec} T$.

Another case when $T_I(A)$ is A-flat is given for $\operatorname{grade}_A I \geq 2$, i.e., $T_I(A) \simeq A$. So one may ask whether $T_I(A)$ is A-flat only if either $\operatorname{cd}_A I \leq 1$ or $\operatorname{grade}_A I \geq 2$. This does not hold by virtue of the following example. Put

$$A = k[x,y,z] \supset I = (x) \cap (y,z) = (xy, xz)A,$$

where x, y, z are indeterminates over a field k. Then $\operatorname{cd}_A I = 2$ and $\operatorname{grade}_A I = 1$. Moreover

$$\operatorname{Ass} A/xyA \cap V(I) = \{xA\}.$$

Because of $1 = x(1/x)$ and $1/x = y/xy = z/xz \in A_{xy} \cap A_{xz}$ it yields $xT = T$. That is, T is Flat over A. In fact, $T = A_x$ as follows from the following observation.

(3.2) Proposition. *Let I, J denote ideals of A. Let $x \in I \cap J$ be a non-zero divisor. Then the following conditions are equivalent:*
i) *$T_I(A) \subseteq T_J(A)$.*
ii) *$xA : (xA : \langle I \rangle) \supseteq xA : (xA : \langle J \rangle)$.*
iii) *$\operatorname{Ass} A/xA \cap V(I) \subseteq \operatorname{Ass} A/xA \cap V(J)$.*

PROOF: First assume $T_I \subseteq T_J$. Then

$$xA : \langle I \rangle = xT_I \cap A \subseteq xT_J \cap A = xA : \langle J \rangle$$

and ii) follows easily. Now assume ii). Let $P \in \operatorname{Ass} A/xA \cap V(I)$, i.e., $P = xA : y \supseteq I$ for an element $y \in A$. Therefore

$$y \in xA : I \subseteq xA : \langle I \rangle \quad \text{and}$$
$$P = xA : y \supseteq xA : (xA : \langle I \rangle).$$

But this means $P \supseteq J$ as easily seen. This proves iii). Now suppose iii). Because $x \in I$ is a regular element one may write any element of T_I in the form $r/x^n, r \in A$. Thus

$$I \subseteq \operatorname{Rad}(x^n A : r).$$

Now choose $\mathfrak{p}$ a minimal prime ideal of $x^n A : r$. Then

$$\mathfrak{p} \in \operatorname{Ass} A/xA \cap V(I) \subseteq \operatorname{Ass} A/xA \cap V(J)$$

by the assumption. That is, $\operatorname{Rad}(x^n A : r) \supseteq J$. Therefore, there is an integer m such that $J^m \subseteq x^n A : r$, i.e., $J^m (r/x^n) \subseteq A$ and $r/x^n \in T_J$, as required.

The previous result is a slight modification of an argument given by Katz and Ratliff, Jr., see [8], 2.3. As an immediate consequence it follows:

(3.3) Corollary. *Let* x, I, J *be as above.*
a) $A = T_I(A)$ *if and only if* $\operatorname{Ass} A/xA \cap V(I) = \emptyset$.
b) $T_I(A) = T_J(A)$ *if and only if* $\operatorname{Ass} A/xA \cap V(I) = \operatorname{Ass} A/xA \cap V(J)$.

Now let $x \in I$ be a non-zero divisor with $xA = \bigcap_{i=1}^s Q_i$ a reduced primary decomposition and $P_i \supseteq I$ for $i = 1, \ldots, t$ and $P_i \not\supseteq I$ for $i = t+1, \ldots, s$, where $P_i = \operatorname{Rad} Q_i$. Put

$$J = \bigcap_{i=1}^t Q_i \text{ if } t > 0 \text{ and } J = A \text{ otherwise.}$$

Then $T_I(A) = T_J(A)$ because of (3.3) and $J = xA : (xA : \langle I \rangle)$ as easily seen.

(3.4) Theorem. *Let* x, I, A *be as above and* $J = xA : (xA : \langle I \rangle)$. *Then the following conditions are equivalent:*
i) $T_I(A)$ *is a flat* A-*module.*
ii) $cd_A J = 1$.
iii) $T_I(A) = J T_I(A)$.

PROOF: Suppose i). Then $T_I = T_J$ is a flat module over A. By (3.1) $T_J = \mathfrak{p} T_J$ for all $\mathfrak{p} \in \operatorname{Ass} A/xA \cap V(J) = \operatorname{Ass} A/J$. Because J is an ideal of a Noetherian ring $\prod_{i=1}^t P_i^{n_i} \subseteq J$ for sufficiently large integers $n_i, i = 1, \ldots, t$. Whence

$$T_J = \prod_{i=1}^t P_i^{n_i} T_J \subseteq J T_J \subseteq T_J$$

and $cd_A J = 1$ by virtue of (2.1). Now suppose ii). If $J = A, T_I = A$ is a flat A-module. Assume J a proper ideal. Then $T_I = T_J$ is a flat A-module by (2.1). Clearly, the last two conditions are equivalent.

Related to (2.2) we get the following finiteness result:

(3.5) Corollary. *Let* I *denote a regular ideal of* A *such that* $T_I(A)$ *is flat over* A. *Then* $T_I(A)$ *is an* A-*algebra of finite type.*

By virtue of the example in Section 2 the converse of (3.5) does not hold in general. In the next section there are some contributions to this problem in dimension two.

4. Dimension two

Before we handle the two-dimensional case let us recall the Hartshorne-Lichtenbaum vasishing theorem on local cohomology, see [6], Theorem 3.1. Let I denote an ideal of a local ring $(A, \mathfrak{m})$ with $d = \dim A$. Then $H_I^d(A) = 0$ if and only if $\dim \widehat{A}/(I\widehat{A} + \mathfrak{p}) > 0$ for all $\mathfrak{p} \in \operatorname{Ass} \widehat{A}$ with $\dim \widehat{A}/\mathfrak{p} = d$. Here $\widehat{A}$ denotes the completion of A. For a different proof see also [2]. In the case of $\dim A = 2$ it yields a characterization when $T_I(A), \operatorname{ht} I = 1$, is an A-algebra of finite type.

(4.1) Theorem. *Let I denote a regular ideal of heigt one in a two-dimensional local ring $(A, \mathfrak{m})$. Then the following conditions are equivalent:*
i) $\operatorname{cd}_A I = 1$.
ii) $T_I(A)$ *is a flat A-module.*
iii) $\dim \widehat{A}/(I\widehat{A} + \mathfrak{p}) = 1$ *for all two-dimensional ideals $\mathfrak{p} \in \operatorname{Ass} \widehat{A}$.*
iv) $\dim T_I(A) = 1$.
v) $T_I(A)$ *is an A-algebra of finite type.*

PROOF: The equivalence of the first three conditions follows easily by (3.4) and the previous remark on the Hartshorne-Lichtenbaum theorem. Note that $\operatorname{Rad} I = \operatorname{Rad} J$. Now suppose i), i.e., $IT = T, T := T_I(A)$. Let $P \in \operatorname{Spec} T$ and $\mathfrak{p} = P \cap A$. Then $\mathfrak{p} \not\supseteq I$ because P is a proper ideal. But this means $A_\mathfrak{p} = T_\mathfrak{p} = T_P$ and $\dim T = 1$. Conversely, suppose $\dim T = 1$. It is enough to show $T = IT$. Assume IT is a proper ideal with $P \supseteq IT$ a minimal prime ideal. Then $\operatorname{ht} P > 1$, see [10], Corollary, p. 61, in contradiction to iv).

Because of (2.2) it follows: i) $\Longrightarrow$ v). In order to complete the proof it remains to show that v) implies $IT = T$. Suppose IT is a proper ideal of T. Then $I\widehat{T}$ is a proper ideal of $\widehat{T} := T \otimes_A \widehat{A} \simeq T_{I\widehat{A}}(\widehat{A})$ by the faithful flatness of $\widehat{A}$ over A. Moreover, if T is an A-algebra of finite type,

$$T = A[I^{-n}], \quad I^{-n} = \{r \in Q(A) : I^n r \subseteq A\},$$

for some $n \in \mathbb{N}$. But then

$$\widehat{T} = T \otimes_A \widehat{A} \simeq \widehat{A}[(I\widehat{A})^{-n}],$$

i.e., $\widehat{T}$ is an A-algebra of finite type. Thus, without loss of generality we may assume A a complete local ring. Let $N \supseteq IT$ denote a minimal prime ideal of IT. Then $\operatorname{ht} N \geq 2$, as above. Let P denote a minimal prime divisor of T such that $P \subseteq N$ and

$$\operatorname{ht} N = \operatorname{ht} N/P \geq 2.$$

Put $\mathfrak{p} = P \cap A$ and $Q = N \cap A$. Then $\mathfrak{p} \subseteq Q$. Now the inclusion $A \subseteq T$ induces an inclusion

$$A/\mathfrak{p} =: A' \longrightarrow T/P =: T'.$$

Set $Q' = QA$ and $N' = NT'$. For the prime ideals Q' and N' it follows $N' \cap A' = Q'$. We apply the dimension formula in the form of Cohen [3], Theorem 1, i.e.,

$$\operatorname{ht} N' + \operatorname{trd}(T'/N')/(A'/Q') \leq \operatorname{ht} Q' + \operatorname{trd} T'/A'.$$

Because of $Q(A') = A_\mathfrak{p}/\mathfrak{p}A_\mathfrak{p}, Q(T') = TP/PT_P$, and $T_P = A_\mathfrak{p}$, note that P is a minimal prime ideal, it yields

$$2 + \text{trd}(T'/N')/(A'/Q') \leq \text{ht}\, Q' \leq 2.$$

Therefore, equality holds and $Q = \mathfrak{m}$. By [3], it follows that

$$T_N/NT_N \simeq T/N \simeq T'/N'$$

is finitely generated over $A/\mathfrak{m} \simeq A'/Q'$. Because $\mathfrak{m}T_N$ is an NT_N-primary ideal it yields that $T_N/\mathfrak{m}T_N$ is finitely generated over $A/\mathfrak{m}$. On the other hand $\bigcap_{n \geq 1} \mathfrak{m}^n T_N = (0)$. Because A is complete, T_N is finitely generated over A, see [9], p. 212. Because of $T \subseteq T_N$ it follows that T is a finitely generated A-module. But this is a contradiction to ht $I = 1$, see [13], (4.4). That is, $IT = T$ as required.

The Theorem generalizes part of [4], (3.,2), where (4.1) is shown for $(A, \mathfrak{m})$ a local Cohen-Macaulay domain such that A', the integral closure of A, is a finite A-module and $A'_{\mathfrak{m}'}$ is analytically irreducible for all maximal ideals $\mathfrak{m}'$ of A'. Among other things, it is shown that under these additional assumptions the above conditons are equivalent to the Noetherianness of $T_I(A)$. The author does not know whether this holds in general. Next we extend [4], (3.2), to the non-Cohen-Macaulay situation.

(4.2) Theorem. *Let $(A, \mathfrak{m})$ be a two-dimensional local domain such that A' is a finitely generated A-module and $A'_{\mathfrak{m}'}$ is analytically irreducible for all maximal ideals $\mathfrak{m}'$ of A'. For an ideal I of A of height one the following conditions are equivalent:*

i) *$T_I(A)$ is an A-algebra of finite type.*
ii) *$T_I(A)$ is a Noetherian ring.*
iii) *The integral closure $T_I(A)'$ of $T_I(A)$ coincides with its complete integral closure.*
iv) *$T_I(A)' = T_{IA'}(A')$.*
v) *$T_{IA'}(A') = IT_{IA'}(A')$.*
vi) *alt $IA' = 1$.*

PROOF: The equivalence of the first four conditions is shown in [4], (3.2), where in fact the Cohen-Macaulay assumption on A is not used. The equivalence of i) and v) follows by (2.4). Furthermore, the implication v) $\implies$ vi) is shown in (2.5). In order to complete the proof let us show vi) $\implies$ v). By a result of Heinzer [7], A' is a Noetherian ring. Therefore, it is enough to prove that $H^2_{IA'}(A')$ vanishes. But this is a local condition. Whence it is enough to prove that $H^2_{IA'\mathfrak{m}'}(A'_{\mathfrak{m}'}) = 0$ for all maximal ideals $\mathfrak{m}'$ of A'. Because of alt $IA' = 1$, it follows that $V(IA')$ is of pure codimension one. Because $A'_{\mathfrak{m}'}$ is analytically irreducible there is only one analytic branch, i.e., the assumptions of the local Hartshorne-Lichtenbaum theorem for $IA'_{\mathfrak{m}'}$ are satisfied. That proves the claim.

As an application of (4.2) there is a characterization of a local domain A as in (4.2) has finitely generated ideal-transforms for every ideal I of height one. This generalizes [11], Theorem 4', p. 53.

(4.3) Corollary. *With A and I as in (4.2) suppose the Going Down Theorem holds for $A \subseteq A'$. Then $T_I(A)$ is an A-algebra of finite type.*

The proof follows easily because alt $IA' = 1$ for every ideal I of height one under the assumptions on $(A, \mathfrak{m})$.

We conclude with a few examples related to the finiteness of $T_I(A)$ in the case of the assumptions of (4.2). To this end note that $T_I(A)$ is of finite type over A if and only if $T_{IA_\mathfrak{m}}(A_\mathfrak{m})$ is of finite type over $A_\mathfrak{m}$ for all maximal ideals $\mathfrak{m}$ of A, see [1], Theorem 6.

(4.4) Examples. a) Let k denote a field. For $A = k[s^2, s^3, t], s, t$ indeterminates over k, it follows $A' = k[s, t]$. Now the Going Down Theorem holds for $A \subseteq A'$. That is, $T_I(A)$ is an A-algebra of finite type for every ideal I of height one.

b) With k, s, t as above let $A = k[s(s-1), s^2(s-1), t]$. Then $A' = k[s, t]$ and the Going Down Theorem does not hold for $A \subseteq A'$, see [9], p. 33. In fact, for $P = (as-t)A', a \in k^*$, and $\mathfrak{p} = P \cap A$ it follows that alt $\mathfrak{p} A' = 2$. To this end note that $g^{-1}(g(P)) = \{P, Q_1, Q_2\}$ with $Q_1 = (s, t-a)A', Q_2 = (s-1, t)A'$, where g denotes the canonical map $\operatorname{Spec} A' \longrightarrow \operatorname{Spec} A$. Hence, there is no height one prime ideal of A' contained in Q_2 and lies over P. So $T_{\mathfrak{p}}(A)$ is not an A-algebra of finite type.

c) Assume $\operatorname{char} k \neq 2$. Put $A = k[x, y, z]/(2yz + y^2 + xz^2), x, y, z$ indeterminates over k, and $\mathfrak{p} = (x, y)A$. Then $\mathfrak{p}$ is a prime ideal of height one such that $T_{\mathfrak{p}}(A)$ is not an A-algebra of finite type. To this end note that $\widehat{A} = k[[x, y, z]]/P_1 \cap P_2$, where $P_i = (y + (1 + (-1)^i u)z)\widehat{A}, i = 1, 2$. Here u is a unit of $\widehat{A}$ with $u^2 = 1 - x$. Then $\operatorname{Ass} \widehat{A} = \{P_1, P_2\}$ and $\dim \widehat{A}/(P_2, \mathfrak{p}\widehat{A}) = 0$ because $1 + u$ is a unit in $\widehat{A}$.

The above example b) is a slight generalization of an example considered in [4]. The example a) resp. c) is studied in [4] resp. [11] from a different point of view.

References

[1] J. BREWER AND W. HEINZER, *Associated primes of principal ideals*, Duke Math. J. **41** (1974), 1–7.

[2] F. W. CALL AND R. Y. SHARP, *A short proof of the local Lichtenbaum-Hartshorne theorem on the vanishing of local cohomology*, Bull. London Math. Soc. **18** (1986), 261–264.

[3] I. S. COHEN, *Length of prime ideal chains*, Amer. J. Math. **76** (1954), 654–668.

[4] P. M. EAKIN, JR., W. HEINZER, D. KATZ AND L. J. RATLIFF, JR., *Notes on ideal-transforms, Rees rings and Krull rings*, J. of Algebra **110** (1987), 407–419.

[5] A. GROTHENDIECK, *Local cohomology*, Lect. Notes in Math. No. 41, Berlin-Heidelberg-New York, 1970.

[6] R. HARTSHORNE, *Cohomological dimension of algebraic varieties*, Ann. of Math. **88** (1968), 403–450.

[7] W. HEINZER, *On Krull overrings of a Noetherian domain*, Proc. Amer. Math. Soc. **22** (1969), 217–222.

[8] D. KATZ AND L. J. RATLIFF, JR., *Two notes on ideal-transforms*, Math. Proc. Camb. Phil. Soc. **102** (1987), 389–397.

[9] H. MATSUMURA, "Commutative Algebra," 2nd edit., New York, 1980.

[10] M. NAGATA, *A treatise on the 14th problem of Hilbert*, Mem. Coll. Sci. Kyoto Univ. **30** (1956), 57–82.

[11] M. NAGATA, *Lecture on the fourteenth problem of Hilbert*, Tata Inst. Fund. Res., Lect. on Math. No. 31, Bombay, 1965.

[12] F. RICHMAN, *Generalized quotient rings*, Proc. Amer. Math. Soc. **16** (1965), 794–799.

[13] P. SCHENZEL, *Finiteness of relative Rees rings and asymptotic prime divisors*, Math. Nachr. **129** (1986), 123–148.

[14] P. SCHENZEL, *Filtrations and Noetherian symbolic blowup rings*, Proc. Amer. Math. Soc. **102** (1988), 817–822.

Sektion Mathematik der Martin-Luther-Universität Halle-Wittenberg, Postfach, DDR-4010 Halle, German Democratic Republic

Topics in Rees Algebras of Special Ideals

ARON SIMIS*

1. Introduction

Let R be a (commutative, noetherian) ring and let $I \subset R$ be an ideal. The main object envisaged in this talk is the *Rees algebra* of I, namely, the graded R-algebra $\mathcal{R}(I) = R \oplus I \oplus I^2 \oplus \dots$. A related ring is the *associated graded ring* of I, to wit, $\mathrm{gr}_I(R) := \mathcal{R}(I)/I\mathcal{R}(I)$. Roughly, the latter can be viewed as a specialization of the former and, in a vague sense, its description is closer to R than that of $\mathcal{R}(I)$. Also, historically perhaps, for reasons of analysing singularities and their resolutions, it was $\mathrm{gr}_I(R)$ that played a central role. As arithmetical questions – in the sense algebraic geometers mean referring to properties of the underlying coordinate ring – increasingly plagued the scene, the emphasis began leaning towards a deeper understanding of the structure of $\mathcal{R}(I)$.

Two broad questions, along with their variations, seem to be unanswered:

(1) Find reasonable conditions under which $\mathcal{R}(I)$ is normal.
(2) Find natural obstructions to the Cohen-Macaulayness of $\mathcal{R}(I)$.

By definiton, the two phenomena intersect along the well-known property S_2 of Serre. Beside this obvious hunch that the properties ought to have more in common, there is no further general evidence to a stronger relationship between them, except for the one single result due to Hochster to the effect that normal subrings generated by monomials in indeterminates over a field are always Cohen-Macaulay. Despite such a lack of evidence, some of us still believe that in the cases of Rees algebras of ideals in special classes the overlapping ought to be detected in a more precise way.

Some of the results stated in this paper will appear *in totum* in [HSV 4]. Two substantial sections are entirely novel, to wit, Section 4, on Graphael ideals, and Section 5, on monomial ideals in two variables. The remaining parts contain variants of proofs of some results in [BST].

2. Normalitana

In this section, we describe a few general techniques for sorting out normal ideals. As a matter of notation, we will sometimes denote the Rees algebra of the ideal I by $R[It]$ to emphasize the embedding $\mathcal{R}(I) \hookrightarrow R[t]$. In this line, yet another algebra that plays an important role in the theory is the so-called *extended* Rees algebra of the ideal I, defined as $R[It, t^{-1}] \subset R[t, t^{-1}]$.

A first criterion for normality is the following bunch of mutually equivalent statements.

---

*Partially supported by a CNPq grant No. 300662/82/MA

(2.1) Proposition. *Let R be a noetherian normal domain and let $I \subset R$ be an ideal. The following conditions are equivalent:*
(i) *The Rees algebra $R[It]$ is normal.*
(ii) *The ideal $IR[It] \subset R[It]$ is integrally closed.*
(iii) *I is normal.*
(iv) *(R local with maximal ideal m) I is locally normal in the punctured spectrum $\operatorname{Spec} R \setminus \{m\}$ and there exists an element $f \in mR[It]$ such that $R[It]_P$ is normal for every $P \in \operatorname{Ass} R[It]/fR[It]$.*
(v) *The ideal $(t^{-1}) \subset R[It, t^{-1}]$ is integrally closed.*
(vi) *$R[It, t^{-1}]_P$ is normal for every $P \in \operatorname{Ass} \operatorname{gr}_I(R)$.*
(vii) *$R[It, t^{-1}]$ is normal.*

The proof of the equivalences is easy (cf. [HSV 4]).

Application. Let I be the ideal generated by the $n - 1 \times n - 1$ minors of a generic $n \times n$ matrix X (over a field k). We show the conditions in (iv) are presently met. First, there is no harm in localizing at (X). Change notation: $R := k[X]_{(X)}$, $I := I_{(X)}$, etc.. By induction on n, using the *inversion and elementary transformation* trick, we have that I is locally normal outside (X). Set $f := \det(X)$. By Huneke's [Hu 2] the ideal $(ft) \subset R[It]$ is radical, hence we are done.

Another criterion bearing theoretical importance is given in terms of normally divisorial ideals.

(2.2) Definition. An ideal $I \subset R$ is *normally divisorial* if the exceptional ideal $IR[It]$ has a primary decomposition

$$IR[It] = P_1^{(l_1)} \cap \cdots \cap P_r^{(l_r)},$$

where P_i is a height one prime ideal of $R[It]$ and $P_i^{(l_i)}$ stands for its l_i-th symbolic power.

(2.3) Proposition. *Let R be a normal domain and let $I \subset R$ be an ideal. The following conditions are equivalent:*
(i) *I is normally divisorial.*
(ii) *I is normal.*

The implication (i) $\Rightarrow$ (ii) is certainly well-known. The reverse one is the bulk of the Proposition: it can be proven by a technique of passing to the extended Rees algebra of I and, once there, applying reduction theory in one-dimensional local rings. For the details, refer to [HSV 4].

In a different vein, there is also a computationally-oriented version of Serre's criterion. Namely, let J stand for a *presentation ideal* of $\mathcal{R}(I)$, i.e., the kernel of a surjective R-homomorphism

$$R[T] := R[T_1, \ldots T_n] \twoheadrightarrow R[It]$$

defined by the assignment $T_i \mapsto x_i t$ for a given set of generators $x_1, \ldots x_n$ of I. Then, one can state:

(2.4) Proposition. ([BSV]) *Let R be a normal domain and let $I \subset R$ be an ideal. The following conditions are equivalent:*
(i) *$R[It]$ is normal.*
(ii) *The following statements hold:*
   (1) *The ideal $(I, J)R[T]$ has no embedded primes;*

(2) *For every minimal prime $P \subset R[T]$ of $R[T]/(I, J)$, the span of J in the $R[T]_P/P_P$- vector space P_P/P_P^2 has codimension one.*

One may note, in this context, that condition (2) above bends by and large to computational devices. Not so condition (1), which requires formidable détours largely dependent on deeper theoretical results (cf., e.g., [Mor] and [Hu 3]).

The next result is a criterion for normality for a special class of ideals. We need a definition:

(2.5) Definition. Let A be a graded ordinal Hodge algebra on a poset H. An ideal $\Omega \subset H$ is *straightening-closed* if, for every incomparable $h, k \in \Omega$, each standard monomial M appearing in the straightening of h, k contains at least two factors from Ω.

This terminology was designed by Bruns and Vetter [BV]. Earlier appearances of the concept are in [Hu 1] and [EH].

An extension of rings $A \subset B$ is said to be *unramified in codimension one* if for every height one prime ideal $p \subset A$, either pA_p is a prime ideal of A_p or else $pA_p = A_p$.

(2.6) Proposition. *Let R be a graded ordinal Hodge algebra over a base ring B on a poset H and let $I \subset R$ be an ideal generated by a straightening-closed ideal in H. Assume:*
(i) R is a normal domain and the extension $B \subset R$ is unramified in codimension one.
(iii) $\mathrm{Cl}(R_{B\setminus 0})$ is a free $\mathbb{Z}$-module.
Then $\mathcal{R}(I)$ is normal and $\mathrm{Cl}(\mathcal{R}(I)) \simeq \mathrm{Cl}(B) \oplus \mathbb{Z}^{r+s}$, where r is the number of minimal primes of $\mathrm{gr}_I(R)$ whose contraction to R have height at least two and $s := \mathrm{rk}\,\mathrm{Cl}(R_{B\setminus 0})$.

The proof of the proposition is largely based on the methods of [ST], particularly on the so-called *fundamental exact sequence* of divisor class groups of blow-ups. For complete details refer to [BST].

3. The Fitting conditions revisited

Let R be a (noetherian) ring . For most theoretical and computational purposes, a finitely generated R-module E is "given" provided a *free presentation*

$$F \xrightarrow{\varphi} G \longrightarrow E \longrightarrow 0$$

is known. The determinantal ideals $I_t(\varphi)$ of various sizes of φ - which, by abuse, we call *Fitting ideals* of E - detect the behaviour of the local number of generators of E. Thus, e.g., one has:

(3.1) Lemma. *For a prime ideal $\mathbf{p} \subset R$ and an integer $t \geq 0$, one has*

$$I_t(\varphi) \not\subset \mathbf{p} \Leftrightarrow \mu(E_{\mathbf{p}}) \leq \mathrm{rk}(G) - t.$$

3.2) Lemma. *As above, suppose moreover that E has a rank. Then, for a prime ideal $\mathbf{p} \subset R$, one has*
$$E_{\mathbf{p}} \text{ is } R_{\mathbf{p}} - \text{free} \Leftrightarrow I_{\mathrm{rk}(G)-\mathrm{rk}(E)}(\varphi) \not\subset \mathbf{p}.$$

The proofs of these lemmata can be found in [Si]. One can further refine the information on the local number of generators of E by considering certain lower bounds for the heights of the Fitting ideals of E.

(3.3) Proposition. *Let R be an equidimensional catenarian domain and let E be a module having a rank and a free presentation as above. For an integer $k > 0$ (resp. $k = 0$), the following conditions are equivalent:*
(i) $\mathrm{ht}(I_t(\varphi)) \geq \mathrm{rk}(\varphi) - t + 1 - k$ *for* $1 \leq t \leq \mathrm{rk}(\varphi)$ *and equality is attained for at least one value of t (resp.* $\mathrm{ht}(I_t(\varphi)) \geq \mathrm{rk}(\varphi) - t + 1$ *for* $1 \leq t \leq \mathrm{rk}(\varphi)$).
(ii) $\mu(E_\mathbf{p}) \leq \mathrm{ht}(\mathbf{p}) + \mathrm{rk}(E) + k$ *for every prime $\mathbf{p} \subset R$ and equality is attained for at least one prime (resp.* $\mu(E_\mathbf{p}) \leq \mathrm{ht}(\mathbf{p}) + \mathrm{rk}(E)$ *for every* $\mathbf{p} \subset R$).
(iii) $\dim S(E) = \dim R + \mathrm{rk}(E) + k$.

Here as in the sequel, $S(E)$ denotes the symmetric algebra of E. The proof is again not difficult [Si]. We remark that equidimensionality and catenaricity are only needed because of (iii), which may fail otherwise [Va]. Similar conditions can be stated for negative values of k, the main differences being to the effect that in the latter case the import is to the free locus of E and not on $\dim S(E)$ anymore - since the inequality $\dim S(E) \geq \dim R + \mathrm{rk}(E)$ always holds.

The maximum value between 0 and the unique integer k for which any of the conditions of (3.3) holds for the module E is called the *Fitting defect* or the *dimension defect* of E and will be denoted $\mathrm{df}(E)$. The Fitting defect admits a (in principle) computable expression directly in terms of the Fitting ideals, namely

(3.4) Proposition. *Let R be an equidimensional catenarian domain and let E be a module having a rank and a free presentation as above. Then*

$$\mathrm{df}(E) = \max_{1 \leq t \leq \mathrm{rk}(\varphi)} \{ \max\{0, \mathrm{rk}(\varphi) - t + 1 - \mathrm{ht}(I_t(\varphi))\} \}.$$

Once more, this is easily checked from the definitions. A slightly different proof is given in [SV]. In this work we focus on ideals of height at least one - that is, roughly, torsionfree modules of rank one with a definite embedding into R. This makes the theory in their case a lot more special. For example:

(3.5) Proposition. *If $I \subset R$ is an ideal of height at least one then*

$$\dim S(I) = \max \{ \dim R + 1, \dim S(I/I^2) \}.$$

For a direct proof, we refer again to [Si]. Of course, it is also a straightforward consequence of the Huneke-Rossi formula [HuRo]. Observe, in addition, that $\dim S(I/I^2) \geq \mu(I/I^2)$ from the same formula, so if, moreover, R is local (or graded and I homogeneous) then $\dim S(I/I^2) \geq \mu(I)$ (again, this is perfectly straightforward without [HuRo]).

(3.6) Definition. The ideal I is of *Valla type* if

$$\dim S(I) = \max \{ \dim R + 1, \mu(I) \}.$$

This definition pays tribute to G. Valla, who was the first to raise the question as to what ideals are of this kind, being himself responsible for a calculation showing that determinantal ideals are not of Valla type (except for a few easily described cases). Incidentally, we know of no nice formula producing the value of $\mathrm{df}(I)$ for a (generic) determinantal ideal.

An important subclass of the above ideals are the ideals of *linear type* - i.e., for which $S(I) = \mathcal{R}(I)$ - and the ideals of *analytic type* - for which $\mu(I) = \ell(I)$, where ℓ stands for *analytic spread*. Both subclasses satisfy $\mu(I) \leq \dim R$. For an account on these, cf. [HSV 1], [HSV 2], [HSV 3] and [Hu 2]. Easy examples of ideals of Valla type are primary ideals to the maximal ideal of a local ring. Apart from these, no large classes of ideals seem to be known to be of Valla type - see, however, the next section.

4. Graphael ideals

Graphael is named after Rafael Villarreal who attaches to a graph (no *loops* allowed) $\mathcal{G}$ on a vertex set $V = \{X_1, \ldots, X_n\}$ the ideal $I(\mathcal{G})$ of the polynomial ring $R := k[X_1, \ldots, X_n]$ generated by the monomials $X_i X_j$ representing the edges of $\mathcal{G}$.

Recall a few definitions from graph theory:

(4.1) Definition.

   (1) A graph $\mathcal{G}$ on $V = \{X_1, \ldots, X_n\}$ is a *cycle* if, up to reordering of the vertices, its edges are $X_1 X_2, \ldots, X_{n-1} X_n, X_n X_1$ (by abuse, cycles with 3,4,5,etc edges will be called *triangles*, *squares*, *pentagons*, etc., respectively.)
   (2) The *cycle rank* of a graph $\mathcal{G}$, denoted $\mathrm{rk}(\mathcal{G})$, is the maximum number of independent cycles of $\mathcal{G}$.

We will need the following analogue of Euler's formula for spherical polihedra:

(4.2) Proposition. *Let $\mathcal{G}$ be a connected graph with n vertices and q edges. Then* $\mathrm{rk}(\mathcal{G}) = q - n + 1$.

References to this result can be found in [Har].

Here is a first result connecting the above formula and the preceding dimensional-theoretic background of symmetric algebras.

(4.3) Proposition. *Let $I := I(\mathcal{G})$ be the Graphael ideal associated to the connected graph $\mathcal{G}$. Then*
$$\mathrm{df}(I) \geq \max \left\{ 0, \mathrm{rk}(\mathcal{G}) - 2 \right\}.$$

PROOF: The case where $\mathcal{G}$ consists of a single edge is trivial, hence, assume $\mathcal{G}$ has at least two edges. In this case, it is clear that every vertex X_i is an entry of the first syzygy matrix φ of I. Therefore, $\mathrm{ht}\,(I_1(\varphi)) = n = q - 1 - (\mathrm{rk}(\mathcal{G}) - 2)$, by Proposition (4.2). The contention now follows from Proposition (3.4). ∎

(4.4) Corollary. *Let $I := I(\mathcal{G})$ be the Graphael ideal associated to the connected graph $\mathcal{G}$. Then*
$$\dim S(I) \geq \max \left\{ \dim R + 1, \mu(I) \right\}.$$

The above inequality is actually an equality:

(4.5) Theorem. ([Vi]) *A Graphael ideal associated to a connected graph is of Valla type.*

PROOF: According to Proposition (3.3) and Proposition (4.3) we are to show that

$$\mu(I_\mathbf{p}) \leq \mathrm{ht}(\mathbf{p}) + 1 + (\mathrm{rk}(\mathcal{G}) - 2)$$

for every prime $\mathbf{p} \subset R$ (and we may as well assume that $\mathbf{p} \supset I$). Equivalently, by Proposition (4.3), the contention is that

$$\dim R/\mathbf{p} \leq \mu(I) - \mu(I_{\mathbf{p}})$$

for every prime $\mathbf{p} \supset I$.

We proceed by induction on $\dim R = \#V$. There is nothing to prove if $\dim R = 1$, so assume $\dim R \geq 2$. Also assume, as we may, that the removal of the vertex X_n does not disconnect $\mathcal{G}$. Set $V' := V \setminus \{X_n\}$ and, correspondingly, $R' := k[X_1, \ldots, X_{n-1}]$, $I' := I \cap R'$, $\mathbf{p}' := \mathbf{p} \cap R'$, etc.

Set $d := \deg X_n$, the cardinal of the set $E_n := \{X_{j_1} X_n, \ldots, X_{j_d} X_n\}$ of all edges adjacent to X_n. Clearly, $I = (I', E_n)$. For the sake of clarity, we separate the discussion into two cases.

1. $X_n \in \mathbf{p}$.

Here, one must have $\mathbf{p} = (\mathbf{p}', X_n)$. If $X_{j_1}, \ldots, X_{j_d} \in \mathbf{p}$ then, easily, $\mu(I_{\mathbf{p}}) = \mu(I'_{\mathbf{p}'}) + d$. If some $X_{j_i} \notin \mathbf{p}$ then $\mu(I_{\mathbf{p}}) = \mu((I', X_n)_{\mathbf{p}}) = \mu(I'_{\mathbf{p}'}) + 1$.

On the other hand, $\mathrm{ht}(\mathbf{p}) = \mathrm{ht}(\mathbf{p}') + 1$. Therefore, since $d \geq 1$, one gets in any case:

$$
\begin{aligned}
\dim R/\mathbf{p} &= \dim R'/\mathbf{p}' \\
&= \mu(I') - \mu(I'_{\mathbf{p}'}) \\
&\leq \mu(I) - d - (\mu(I_{\mathbf{p}}) - d) \\
&= \mu(I) - \mu(I_{\mathbf{p}}).
\end{aligned}
$$

2. $X_n \notin \mathbf{p}$.

At any rate, $\mathrm{ht}(\mathbf{p}) \geq \mathrm{ht}(\mathbf{p}')$. On the other hand, one clearly has

$$I_{\mathbf{p}} = (I', X_{j_1}, \ldots, X_{j_d})_{\mathbf{p}}.$$

Now, the crucial point is to observe that, since every X_{j_i} is adjacent to at least one vertex other than X_n and the corresponding edge is an element of I', it follows that $\mu(I_{\mathbf{p}}) \leq \mu(I'_{\mathbf{p}'})$.

The inductive hypothesis applies here again to yield the desired contention. ∎

(4.6) **Corollary.** *If $I \subset R$ is the Graphael ideal of a connected graph $\mathcal{G}$, then*

$$
\dim S(I) = \begin{cases}
\dim R + 1 & \text{if } \mathrm{rk}(\mathcal{G}) \leq 2 \\
\dim R - 1 + \mathrm{rk}(\mathcal{G}) & \text{if } \mathrm{rk}(\mathcal{G}) > 2.
\end{cases}
$$

(4.7) **Example.** Graphs with two vertex coverings.

The corresponding Graphael ideal is of the form

$$I := (X_1, \ldots, X_n) \cap (Y_1, \ldots, Y_m)$$

for some n, m. If $n, m \geq 2$ then the graph has at least two independent cycles. By the preceding results, $\dim S(I) = mn$ and $\mathrm{df}(I) = mn - (m + n) - 1 = (m - 1)(n - 1)$. It also follows that the height of the presentation ideal of $S(I)$ is $m + n$. This should be compared to the results in [BST], where it is shown that the presentation ideal of the *extended* Rees algebra of I is generated by the 2×2 minors of a generic matrix. Incidentally, [BST] contains the result to the effect that the Rees algebra of I is Cohen-Macaulay, which gives some support to a question posed by Villarreal [Vi]. Since [BST] uses Hodge algebras methods, it seems natural to pose the following

Problem. (i) Characterize the Graphael ideals whose generators form an ideal in the poset of some (naturally defined) Hodge algebra structure on $R = k[X]$.

(ii) For a Graphael ideal as in (i) is there any obvious relationship between its graph and the graph of the underlying poset of the Hodge structure?

Here is a modest contribution to the this problem.

(4.8) Theorem. *Let* $X := X_1, \ldots, X_m; Y := Y_1, \ldots, Y_n; Z := Z_1, \ldots, Z_p; \ldots$ *be a finite collection of (mutually independent) sets of indeterminates over* k. *Set* $R := k[X, Y, Z, \ldots]$, $V := \{X, Y, Z, \ldots\}$ *and let* $\mathcal{G}$ *be the graph whose vertex set is* V *and whose edge set is* $E := \{X_i Y_j, X_i Z_k, Y_j Z_k, \ldots\}$. *Let* $I := I(\mathcal{G})$ *denote the corresponding Graphael ideal. Then:*

(i) R *has a structure of graded ordinal Hodge algebra on a poset* $H \subset (X, Y, Z, \ldots)$ *such that the generators of* I *form a straightening-closed ideal in* H.

(ii) *The Rees algebra of* I *(and, consequently, the associated graded ring of* I) *is Cohen-Macaulay.*

(iii) *The associated graded ring of* I *is Gorenstein if and only* $\#X = \#Y = \#Z = \ldots$.

Let further $m(I)$ *denote the number of (mutually independent) sets of indeterminates in* V. *Then:*

(iv) $m(I)$ *is the number of minimal primes of* R/I.

(v) *The Rees algebra* $\mathcal{R}(I)$ *is normal with divisor class group* $\mathrm{Cl}(\mathcal{R}(I)) \simeq \mathbb{Z}^{m(I)}$.

(vi) *The ideal* I *is of linear type in exactly one of the following cases: 1)* $m(I) = 1$ *(totally disconnected graph); 2)* $m(I) = 2$ *and, say,* $\#X = 1$; *3)* $m(I) = 3$ *and* $\#X = \#Y = \#Z = 1$ *(triangle).*

PROOF: (i) Set $H := V \cup E \subset R$. Endow H with a structure of poset as follows:

$$X_1 \leq \cdots \leq X_m; \quad Y_1 \leq \cdots \leq Y_n; \quad Z_1 \leq \cdots \leq Z_p; \ldots$$
$$X_{i_1} Y_j < X_{i_2} \qquad X_{i_1} Z_k < X_{i_2} \qquad Y_{j_1} Z_k < Y_{j_2}; \ldots$$
$$X_i Y_{j_1} < Y_{j_2} \qquad X_i Z_{k_1} < Z_{k_2} \qquad Y_j Z_{k_1} < Z_{k_2}; \ldots$$
$$X_{i_1} Y_{j_1} \leq X_{i_2} Y_{j_2} \quad X_{i_1} Z_{k_1} \leq X_{i_2} Z_{k_2} \quad Y_{j_1} Z_{k_1} \leq Y_{j_2} Z_{k_2}; \ldots$$
$$X_{i_1} Y_j \leq X_{i_2} Z_k \quad X_i Y_{j_1} \leq Y_{j_2} Z_k \quad X_i Z_{k_1} \leq Y_j Z_{k_1}; \ldots.$$

The convention about the indices above is that i, j, k are arbitrary and $i_1 \leq i_2, j_1 \leq j_2, k_1 \leq k_2$. We stress the point to the effect that, although two variables from different sets are incomparable, there is a definite choice of *enumeration* of these sets. It is not difficult to see that H is built out of poset blocks, each identifiable with the poset of entries of a generic matrix with greatest element removed.

Now, in order to prove that R is an ordinal Hodge algebra on H, it is sufficient to show that the standard monomials on H are linearly independent. But, an arbitrary (ordinary) multimonomial in $X, Y, Z, \ldots$, say,

$$M := X_1^{r_1} \ldots X_m^{r_m} Y_1^{s_1} \ldots Y_n^{s_n} Z_1^{t_1} \ldots Z_p^{t_p} \ldots$$

can be written as

$$(X_1 Y_1 Z_1 \dots)^{min(r_1, s_1, t_1, \dots)} . M'$$

for some factor M' and further, the standard monomial representation of M has to start with the factor $(X_1 Y_1 Z_1 \dots)^{min(r_1, s_1, t_1, \dots)}$. The rest follows by induction.

One can write explicitly all straightenings of pairs of incomparable elements of H. In particular, the poset ideal $E \subset H$ is then seen to be straightening-closed. Since $I = (E)R$, this proves our first contention.

(ii) The proof of this part is an immediate application of [BV, (9.12)] (cf. also [EH]), provided one checks the following statements: H is a wonderful poset, E is a self-covering ideal and contains all the minimal elements of H. Al these statements are easily verified from the definitions.

(iii) Since I is generated by a poset ideal in a graded ordinal Hodge algebra, the associated graded ring $G := gr_I(R)$ is reduced [BV, (9.8)]. As R is regular, it follows from [HSV 4, (5.2)] that G is Gorenstein if and only if I is unmixed. On the other hand, say $V = V_1 \sqcup \cdots \sqcup V_s$ is the partition of the vertex set into the given collection of nonoverlapping subsets. A moment of reflection will convince us that the minimal primes of R/I are

$$(V_1, \dots, V_{s-1}, \widehat{V_s})R, \quad (V_1, \dots, \widehat{V_{s-1}}, V_s)R \dots (\widehat{V_1}, V_2, \dots, V_s)R.$$

A trivial combinatorial argument then shows that I is unmixed if and only if $\#V_1 = \#V_2 = \cdots = \#V_s$.

(iv) This part has been essentially taken care of in the proof of (iii).

(v) It is well known that, under the circumstances that R is normal and G is reduced, the Rees algebra $\mathcal{R}(I)$ is normal. The calculation of the divisor class group is a consequence of the methods developed in [ST] (cf. also [HuSV]).

(vi) First, we show that if $m(I) \geq 4$ then I can't be of linear type. Indeed, pick four vertices X_1, Y_1, Z_1, W_1 from mutually nonoverlapping sets. The edges $X_1 Y_1$, $X_1 W_1$, $Y_1 Z_1$, $Z_1 W_1$ yield a Rees relation which is not generated by the linear ones.

Next, the same vein of argument shows that, for $m(I) \geq 2$, at most *one* of the vertex subsets V_i can have more than one single element. Therefore, the stated conditions are indeed necessary. Clearly, the cases listed are easily seen to be of linear type directly. ∎

(4.9) Remark. The unmixed case of Theorem (4.8) is a *regular* graph. Whether regularity is a natural obstruction to a Hodge structure in general, remains to be answered. Note, e.g., the two regular 3-graphs with 6 vertices:

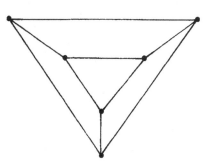 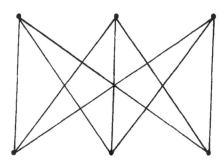

Note that only the second one fits in the class under consideration. As a matter of fact, the heights of the corresponding Graphael ideals are different (4 and 3, respectively), which leads us to guess that Hodge-like unmixed Graphael ideals form a special component of the totality of regular Graphael ideals.

5. Monomial ideals in two variables

The Rees algebra of an m-primary monomial ideal $I \subset k[x, y]$, $m := (x, y)$, is as yet not completely understood from the viewpoint of normality and Cohen-Macaulayness. True, there is the *convex hull* criterion for normality, which holds quite generally for any monomial ideal ([KM], [LT]), but it is hardly the case that such a criterion can be immediately transposed in terms of the initial data so as to gain in conceptuality and transparency. Efforts towards the understanding of the problem have been employed by various authors, cf., e.g., [HuS], [BSV], [HSV 4].

5.1. Three-generated ideals. By far this seems to be the best understood case. Pieces of the following result may appear scattered in the literature, but to our knowledge no complete statements have yet been formulated elsewhere.

(5.1.1) Theorem. *Let* $I := (x^a, x^c y^d, y^b) \subset R := k[x, y]$, *with* $a > c \geq 1, b > d \geq 1$. *Then:*
(i) $\mathcal{R}(I)$ *is Cohen-Macaulay if and only if either* $a \geq 2c, b \geq 2d$ *or* $a \leq 2c, b \leq 2d$.
(ii) $\mathcal{R}(I)$ *is normal if and only if one of the following cases takes place*

    (1) $c = 1, a = 2, b \geq 2d - 1$;
    (2) $d = 1, b = 2, a \geq 2c - 1$.

PROOF: Map $R[T, U, V] \longrightarrow \mathcal{R}(I)$ by sending $T \mapsto x^a, U \mapsto x^c y^d, V \mapsto y^b$, and let J stand fro the corresponding kernel. We claim that

$$J = (y^d T - x^{a-c} U, x^c V - y^{b-d} U, \begin{cases} TV - x^{a-2c} y^{b-2d} U^2) \text{ if } a \geq 2c, b \geq 2d \\ U^2 - x^{2c-a} y^{2d-b} TV) \text{ if } a \leq 2c, b \leq 2d \end{cases}$$

Consider first the case where $a \geq 2c, b \geq 2d$. Clearly, the displayed polynomials belong to J and generate an ideal of height two. Therefore, it suffices to show that the ideal J' they generate is prime. For this, observe that these polynomials are the 2×2 minors of the matrix

$$\begin{pmatrix} T & y^{b-2d} U & x^c \\ x^{a-2c} U & V & y^d \end{pmatrix},$$

hence J' is an unmixed ideal. It is straightforward to check that T is a non-zero-divisor on J', so it remains to verify that $J'[1/T] \subset R[T, U, V][1/T]$ is prime, which in turn amounts to checking that the polynomial $y^d - (1/T) x^{a-c} U$ is irreducible in $k(x, y)[T, 1/T, U]$. The last statement is clear.

We have thus proved that J is determinantal, hence Cohen-Macaulay. The case where $a \leq 2c, b \leq 2d$ is entirely similar. Next, conversely, assume (say) $a > 2c$ and $b < 2d$. We easily see that

$$J' := y^d T - x^{a-c} U, x^c V - {}^{b-d} U, x^{a-2c} U^2 - y^{2d-b} TV \subset J.$$

But here, $J' \subset mR[T, U, V]$, hence necessarily, $J' \neq J$ since the analytic spread of I is two. On the other hand, I admits in this case no other *fresh* quadratic relations in T, U, V (only cubic or higher). It follows, by reasons of degree, that J is not determinantal, hence not Cohen-Macaulay. This proves the first statement.

In order to prove (ii), we may assume at the outset one of the cases $a \geq 2c, b \geq 2d$ or $a \leq 2c, b \leq 2d$ [BSV, (7.3)]. Consider the first of these cases. By the criterion of Proposition (2.4), we ought to figure out the rank of the following jacobian matrix

$$
\begin{pmatrix}
-(a-c)x^{a-c-1}U & cx^{c-1}V & (a-2c)x^{a-2c-1}y^{b-2d}U^2 \\
dy^{d-1}T & (b-d)y^{b-d-1}U & (b-2d)y^{b-2d-1}x^{a-2c}U^2 \\
y^d & 0 & (b-2d)y^{b-2d-1}x^{a-2c}U^2 \\
-x^{a-c} & y^{b-d} & 2x^{a-2c}y^{b-2d}U \\
0 & x^c & T
\end{pmatrix},
$$

modulo the associated primes of I, J. The following possibilities arise:

(1) $a = 2c, b > 2d$.

The associated primes of (I, J) are, in this case, the ideals (x, y, T) and (x, y, V). There is a complete symetry here, so we'll consider only the first of them. If $c = 1$,

$$
\det \begin{pmatrix} V & (a-2)y^{b-2d}U^2 \\ 0 & V \end{pmatrix} \neq 0.
$$

so the matrix has the correct rank for normality along (x, y, T). Assume then $c > 1$, so that $a > c + 1$ as $a = 2c$. Since $b > 2d$, we must have $b > d + 1$ too, so the matrix has actually one single nonzero entry.

(2) $a = 2c, b = 2d$.

There is one single associated prime in this case, namely, $(x, y, TV - U^2)$. Since $c = 1$ is subsumed in the preceding case, we're left with $c > 1$. The possibility that, simultaneously, $d > 1$ is easily ruled out modulo m. Therefore, $d = 1$ must be the case, when it is seen that

$$
\det \begin{pmatrix} T & 0 \\ 0 & T \end{pmatrix} \neq 0.
$$

(3) $a > 2c$.

By symmetry, this has already been considered. Summing up, in the set-up under consideration, namely, $a \geq 2c, b \geq 2d$, $\mathcal{R}(I)$ is normal exactly in one of the cases $c = 1, a = 2$ or $d = 1, b = 2$.

Finally, consider the case where $a \leq 2c, b < 2d$. Here, (x, y, U) is the only associated prime of (I, J). Assume first that $a < 2c$, in which case we must have $c > 1$ and $d > 1$. It is then easy to see that, modulo this prime, the jacobian matrix is the zero matrix. So, let $a = 2c$. In this case, the matrix reduces to the form

$$
\begin{pmatrix}
0 & -cx^{c-1}V & 0 \\
0 & 0 & -(b-2d)y^{2d-b-1}TV \\
0 & 0 & 0 \\
0 & 0 & 0 \\
0 & 0 & 0
\end{pmatrix}.
$$

This matrix has rank two if and only if $c = 1$ (hence $a = 2$) and $b = 2d - 1$. This completes the second case and finishes the proof of (ii). ∎

The conditions as above under which Cohen-Macaulayness materializes suggest introducing the following notion:

(5.1.2) Definition. Let $(a,b) \in \mathbb{N} \times \mathbb{N}$. The (a,b)-*complement* of a pair $(c,d) \in \mathbb{N} \times \mathbb{N}$ such that $c < a, d < b$ is the pair $(a-c, b-d)$.

If needed, whenever the reference pair (a,b) is understood, we will denote the complement of (c,d) by $(c,d)^{\perp}$. Accordingly, we speak of the *complement* of the monomial ideal $I := (x^a, x^c y^d, y^d)$ as being the ideal $I^{\perp} := (x^a, x^{a-c} y^{b-d}, y^b)$. As a consequence of part (i) of the theorem, we have:

(5.1.3) Corollary. *If $\mathcal{R}(I)$ is Cohen-Macaulay then $\mathcal{R}(I^{\perp})$ is Cohen-Macaulay.*

(5.1.4) Remark. Given the pair (a,b), we may consider the rhombus determined by the lattice points $(a,0), (1,1), (0,b), (a-1, b-1)$. One can verify that taking complements is a central symmetry with respect to the meeting point of the diagonals of the rhombus. From this it follows that (5.1.3) would in turn cut roughly by half the verification of the Cohen-Macaulay cases in the theorem.

5.2 A special instance: exponents in arithmetic progression. There are surely many ways in which the preceding results could be extended. We will present one which is fairly simple yet retains the main features of a more general classification.

Consider an (x,y)-primary ideal generated by monomials,

$$I := (x^{c_0}, x^{c_1} y^{d_1}, \dots, x^{c_{n-1}} y^{d_{n-1}}, y^{d_n})$$

where the exponent sequences $\{c_0, c_1, \dots, c_{n-1}\}$ and $\{d_1, \dots, d_n\}$, $(n \geq 2)$, satisfy the following conditions:

(1) $c_0 \geq 2c_1$ (resp. $d_n \geq 2d_{n-1}$);
(2) $c_i = (n-i)c_{n-1}$ for $i = 1, \dots, n-1$ (resp. $d_j = jd_1$ for $j = 1, \dots, n-1$).

For reasons that will become clear later, we call these ideals *semi-catalectican*. We proceed to the main result concerning such ideals.

(5.2.1) Theorem. *Let $I \subset R := k[x,y]$ as above be a semi-catalectican ideal. Then:*
(i) *$\mathcal{R}(I)$ is Cohen-Macaulay.*
(ii) *Setting $a := c_0 + c_2 - 2c_1, b := d_n + d_{n-2} - 2d_{n-1}, c := c_1 - c_2(= c_{n-1}), d := d_{n-1} - d_{n-2}(= d_1), \mathcal{R}(I)$ is normal in exactly one of the following cases:*

(1) $a = b = 0$; $c = 1$ *or* $d = 1$
(2) $a = 0, b \geq 1$; $c = 1$
(3) $a \geq 1, b = 0$; $d = 1$
(4) $a \geq 1, b \geq 1$; $c = d = 1$.

PROOF: (i) The proof is analogous to that of the three-generators case. As before, map $R[T_0, \dots, T_n] \longrightarrow \mathcal{R}(I)$ by sending $T_0 \mapsto x^{c_0}, T_n \mapsto y^{d_n}$ and $T_i \mapsto x^{c_i} y^{d_i}$ for $i = 1, \dots, n-1$. From the very definition of the exponent sequences it easily follows that the kernel J contains the ideal $I_2(M)$ generated by the 2×2 minors of the matrix

$$M := \begin{pmatrix} T_0 & T_1 & \cdots & T_{n-2} & y^{d_n + d_{n-2} - 2d_{n-1}} T_{n-1} & x^{c_{n-1}} \\ x^{c_0 + c_2 - 2c_1} T_1 & T_2 & \cdots & T_n - 1 & T_n & y^{d_1} \end{pmatrix}$$

We first claim that $I_2(M)$ is a perfect ideal. For this. it will suffice to show that its height is largest possible, namely, n. But, the 2×2 minors of the submatrix

$$\begin{pmatrix} T_1 & T_2 & \cdots & T_{n-2} & x^{c_{n-1}} \\ T_2 & T_3 & \cdots & T_{n-1} & y^{d_1} \end{pmatrix}$$

clearly generate an unmixed ideal of height $n-2$ and, moreover, a calculation yields that the minors $T_0 T_2 - x^{c_0+c_2-2c_1} T_1^2, T_n T_1 - y^{d_n+d_{n-2}-2d_{n-1}} T_2 T_{n-1}$ form a regular sequence modulo this ideal. Therefore, $I_2(M)$ has the required height and, to show equality $I_2(M) = J$, it will suffice to prove that $I_2(M)$ is a prime ideal. Note $I_2(M)$ is anyway unmixed since it is perfect. We now claim that T_0 is a nonzerodivisor on $I_2(M)$. If not, let $P \supset I_2(M)$ be an associated prime containing T_0. From the form of the generators, one ought to have either $x \in P$ or $T_1 \in P$. First $T_1 \in P$. In this case, $T_2^2 = (T_2^2 - T_1 T_3) + T_1 T_3 \in P$, hence $T_2, T_3, \ldots, T_{n-2} \in P$. But then, also $T_{n-1} \in P$ or $y \in P$, in which case P would have height $n+1$, contradicting unmixedness. In a similar vein, let $x \in P$. Then $y \in P$ or $T_{n-1} \in P$. In the first case, P contains x, y, T_0 and the 2×2 of the catalectican matrix

$$\begin{pmatrix} T_1 & T_2 & \ldots & T_{n-2} \\ T_2 & T_3 & \ldots & T_{n-1} \end{pmatrix}$$

thus contributing height n already. Next, if $d_n + d_{n-2} = 2d_{n-1}$ then actually the 2×2 of the matrix

$$T := \begin{pmatrix} T_1 & \ldots & T_{n-1} \\ T_2 & \ldots & T_n \end{pmatrix}$$

are contained in P, which is a contradiction. Else, $T_{n-2} \in P$ or $T_n \in P$, again impossible by a height counting. The remaining possibility, $T_{n-1} \in P$, is similarly ruled out.

Having shown T_0 is a nonzerodivisor on $I_2(M)$, the usual procedure of inverting a variable and effecting elementary transformations easily yields that $I_2(M)$ is prime.

(ii) We will use the criterion of (2.4). First, note that condition (1) in that criterion is automatically satisfied as $\mathcal{R}(I)$ - hence $\mathrm{gr}_I(R)$ - is Cohen-Macaulay. Next, we need a neat description of the minimal primes of (I, J). The following bit of notation will be useful: T_i^j will stand for the 2-row submatrix of the catalectican matrix T with columns ith through jth (indexing starts from 0). The list of the required primes is as follows:

(1) $Q := (x, y, I_2(T_0^{n-1}))$ provided $c_0 + c_2 = 2c_1, d_n + d_{n-2} = 2d_{n-1}$.
(2) $Q_1 := (x, y, T_0, \ldots, T_{n-2})$ and $Q_2 := (x, y, T_n, I_2(T_0^{n-2}))$ provided $c_0 + c_2 = 2c_1, d_n + d_{n-2} > 2d_{n-1}$.
(3) $Q_1' := (x, y, T_2, \ldots, T_n)$ and $Q_2' := (x, y, T_0, I_2(T_1^{n-1}))$ provided $c_0 + c_2 > 2c_1, d_n + d_{n-2} = 2d_{n-1}$.
(4) Q_1 as in (2), Q_1' as in (3) and $Q_3'' := (x, y, T_0, T_n, I_2(T_1^{n-2}))$ provided $c_0 + c_2 > 2c_1, d_n + d_{n-2} > 2d_{n-1}$.

The verification is a bit lengthy but straightforward, so we leave out the details. We now proceed to checking the ranks of (I, J) along each one of these primes.

(1) We're assuming that

$$\begin{cases} c_0 + c_2 = 2c_1 & \text{(i.e., } c_0 = nc_{n-1}) \\ d_n + d_{n-2} = 2d_{n-1} & \text{(i.e., } d_n = nd_1). \end{cases}$$

We set $c := c_{n-1}, d := d_1$. Let $\Delta_1, \ldots, \Delta_{n-1}$ stand for a maximal regular sequence of minors in $I_2(T_0^{n-1})$. Then $x, y, \Delta_1, \ldots, \Delta_{n-1}$ generate the ideal Q_Q and also form a basis of the vector space Q_Q/Q_Q^2. The matrix in this basis corresponding to the vector subspace spanned by the elements in J is (apart from nonzero integer factors)

$$
\begin{pmatrix}
0 & 0 & \cdots & 0 & 0 & \cdots & 0 & x^{c-1}T_1 & x^{c-1}T_2 & \cdots & x^{c-1}T_n \\
0 & 0 & \cdots & 0 & 0 & \cdots & 0 & -y^{d-1}T_0 & -y^{d-1}T_1 & \cdots & -y^{d-1}T_{n-1} \\
1 & 0 & \cdots & 0 & * & \cdots & * & 0 & 0 & \cdots & 0 \\
0 & 1 & \cdots & 0 & * & \cdots & * & 0 & 0 & \cdots & 0 \\
\vdots & \vdots & \ddots & \vdots & \vdots & \cdots & \vdots & \vdots & \vdots & \ddots & \vdots \\
0 & 0 & \cdots & 1 & * & \cdots & * & 0 & 0 & \cdots & 0
\end{pmatrix},
$$

where the first block comes from the generators $\Delta_1, \ldots, \Delta_{n-1}$, and the second and third block come, respectively, from the remaining minors not involving x, y and those involving these variables. Clearly, the columns of the second block are linearly dependent upon the first block. Therefore, the matrix has the required rank n (i.e., a nonzero $n \times n$ minor) if and only if $c = 1$ or $d = 1$.

(2) The assumption is that

$$
\begin{cases}
c_0 + c_2 = 2c_1 \\
d_n + d_{n-2} > 2d_{n-1}.
\end{cases}
$$

Set $b := d_n + d_{n-2} - 2d_{n-1}$. First consider the prime Q_1. One sees that $x, y, T_0, \ldots, T_{n-2}$ form a basis of $Q_{1Q_1}/Q_{1Q_1}^2$ and the matrix corresponding to the generators of J in this basis is

$$
\begin{pmatrix}
0 & \cdots & 0 & 0 & 0 & \cdots & 0 & 0 & 0 & \cdots & 0 & 0 & 0 \cdots 0 & x^{c-1}T_{n-1} & x^{c-1}T_n \\
0 & \cdots & 0 & 0 & 0 & \cdots & 0 & 0 & 0 & \cdots & 0 & y^{b-1}T_{n-1}^2 & 0 \cdots 0 & 0 & y^{b+d-1}T_{n-1} \\
0 & \cdots & 0 & T_{n-1} & 0 & \cdots & 0 & T_n & & & & & \\
0 & \cdots & 0 & 0 & T_{n-1} & \cdots & 0 & 0 & T_n & & & & \\
\vdots & \ddots & \vdots & \vdots & & \vdots & \ddots & \vdots & & \vdots & \ddots & & \\
0 & \cdots & 0 & 0 & 0 & & \cdots & T_{n-1} & 0 & 0 & \cdots & T_n & & \\
0 & \cdots & 0 & 0 & 0 & & \cdots & 0 & 0 & 0 & \cdots & 0 & T_n &
\end{pmatrix}.
$$

Here, the first block comes from the minors with columns $i, i+1$ for $1 \le i \le n-3$, the second from minors with columns $i, n-1$ for $1 \le i \le n-2$, the third from those with columns i, n for $1 \le i \le n-1$ and the fourth block from the remaining ones (linear in the T-variables). Again we need a nonzero $n \times n$ minor. As $b + d - 1 \ge 1$, a close inepection of the matrix reveals that such a minor exists if and only if $c = 1$.

Next consider the prime Q_2. A basis for $Q_{2Q_2}/Q_{2Q_2}^2$ is $x, y, T_n, \Delta_1, \ldots, \Delta_{n-2}$, by a choice of a maximal regular sequence of minors in $I_2(T_0^{n-2})$. Relative to this basis, the matrix is now

$$
\begin{pmatrix}
0 & 0 & \cdots & 0 & 0 & \cdots & 0 & 0 & \cdots & 0 & x^{c-1}T_1 & \cdots & x^{c-1}T_{n-1} & x^{c-1}T_n \\
0 & 0 & \cdots & 0 & 0 & \cdots & 0 & y^{b-1}T_1T_{n-1} & \cdots & y^{b-1}T_{n-1}^2 & y^{d-1}T_0 & \cdots & y^{d-1}T_{n-2} & y^{b+d-1}T_{n-1} \\
0 & 0 & \cdots & 0 & 0 & \cdots & 0 & T_0 & & \cdots & T_{n-2} & & & \\
1 & 0 & \cdots & 0 & & & & & & & & & \\
0 & 1 & \cdots & 0 & & & & & & & & & \\
\vdots & \vdots & \ddots & \vdots & & & & & & & & & \\
0 & 0 & \cdots & 1 & & & & & & & & &
\end{pmatrix},
$$

where it is understood that blanks are zeroes. In order to obtain rank n, the elligible minors are of the form

$$
\begin{pmatrix}
0 & 0 & \cdots & 0 & 0 & x^{c-1}T_j \\
0 & 0 & \cdots & 0 & T_i & 0 \\
1 & 0 & \cdots & 0 & & \\
0 & 1 & \cdots & 0 & & \\
\vdots & \vdots & \ddots & \vdots & & \\
0 & 0 & \cdots & 1 & &
\end{pmatrix}
\quad \text{or} \quad
\begin{pmatrix}
0 & 0 & \cdots & 0 & y^{b-1}T_{i+1}T_{n-1} & y^{d-1}T_k \\
0 & 0 & \cdots & 0 & T_i & 0 \\
1 & 0 & \cdots & 0 & & \\
0 & 1 & \cdots & 0 & & \\
\vdots & \vdots & \ddots & \vdots & & \\
0 & 0 & \cdots & 1 & &
\end{pmatrix},
$$

where $0 \leq i, k \leq n-2, 1 \leq j \leq n$. Clearly, one of these minors is nonzero if and only if $c = 1$ or $d = 1$.

(3) This case is completely symmetric to (2).

(4) We are assuming that $a > 0$ and $b > 0$. Along the primes Q_1' and Q_2', the discussion was essentially worked out above (note the matrix is now even *worse* for the purpose of getting a nonzero $n \times n$ minor). The outcome is that the rank along these primes is n if and only if $c = 1$ *and* $d = 1$. It remains to consider the prime Q_3''. A basis is $x, y, T_0, T_n, \Delta_2, \ldots, \Delta_{n-2}$. Again, separating the generators of J into convenient blocks, the matrix looks like

$$\begin{pmatrix}
0\,0 & \cdots & 0\,0 & \cdots & 0 & 0 & \cdots & 0 & T_1^2\, T_1 T_2 & \cdots & T_1 T_{n-2}\, 0 & x^{c-1} T_2 & \cdots & x^{c-1} T_{n-1}\, 0 \\
0\,0 & \cdots & 0\,0 & \cdots & 0\, y^{b-1} T_2 T_{n-1} & \cdots & y^{b-1} T_{n-1}^2 & 0 & 0 & \cdots & 0 & 0\, y^{d-1} T_1 & \cdots & y^{d-1} T_{n-2} \\
0\,0 & \cdots & 0\,0 & \cdots & 0 & 0 & \cdots & 0 & T_2 & T_3 & \cdots & T_{n-1} \\
0\,0 & \cdots & 0\,0 & \cdots & 0 & T_1 & \cdots & T_{n-2} \\
1\,0 & \cdots & 0 \\
0\,1 & \cdots & 0 \\
\vdots\, \vdots & \ddots & \vdots \\
0\,0 & \cdots & 1
\end{pmatrix}.$$

The most *favourable* $n \times n$ minors are

$$\begin{pmatrix}
0\,0 & \cdots & 0\, y^{b-1} T_{i+1} T_{n-1} & 0 & y^{d-1} T_k \\
0\,0 & \cdots & 0 & 0 & T_j & 0 \\
0\,0 & \cdots & 0 & T_i & 0 & 0 \\
1 \\
0\,1 \\
\vdots\,\vdots & \ddots \\
0\,0 & 0 & 1
\end{pmatrix} \quad \text{and} \quad \begin{pmatrix}
0\,0 & \cdots & 0\,0 & T_1 T_{j-1} & x^{c-1} T_k \\
0\,0 & \cdots & 0\,0 & T_j & 0 \\
0\,0 & \cdots & 0\, T_i & 0 & 0 \\
1 \\
0\,1 \\
\vdots\,\vdots & \ddots \\
0\,0 & 0 & 1
\end{pmatrix}.$$

One sees that one of these is nonzero if and only if $c = 1$ or $d = 1$. We observe that the remaining elligible minors still require these conditions plus $a = 1$ or $b = 1$. Since one needs regularity along *all* primes in this class, one ought to have $c = 1$ and $d = 1$. ∎

6. Virtual maximal minors

6.1. Full ideals of maximal minors. By a *full* ideal of maximal minors we mean an ideal

$$I := I_s(X)/I_{s+1}(X) \subset R := B[X]/I_{s+1}(X)$$

where X is an $m \times n$ generic matrix and $1 \leq s \leq min\{m, n\}$. Note that the ordinary maximal minors are obtained when $s = min\{m, n\}$.

(6.1.1) Proposition. *Let* $I \subset R$ *be a full ideal of maximal minors and assume the base ring* B *is a Cohen-Macaulay normal domain. Then:*
(i) $gr_I(R)$ *is a normal domain.*
(ii) $\mathcal{R}(I)$ *is normal and* $Cl(\mathcal{R}(I)) \simeq Cl(B) \oplus \mathbb{Z}^r$, *where*

$$r = \begin{cases} 0, & \text{for } s = 1 = m = n \\ 1, & \text{for } s = 1 = min\{m, n\} < max\{m, n\} \\ 2, & \text{else.} \end{cases}$$

PROOF: (i)Since $R/I \simeq B[X]/I_s(X)$ is a normal domain [HE] and $\mathrm{gr}_I(R)$ is Cohen-Macaulay, in order to show that $\mathrm{gr}_I(R)$ is a normal domain, it is enough to show a certain growth pattern for the local analytic spreads of I, to wit [EH, Proposition 3.2]

$$\ell(I_p) \leq \max\{\,\mathrm{ht}(I_p)\,,\,\mathrm{ht}(p) - 2\}$$

for all primes $p \supset I$.

We induct on s.

Let $s > 1$ and assume first that $p \supset XR$. We may clearly suppose that p is a homogeneous ideal. Then, $p = (p \cap B, X)R$, hence by applying the base change $B \longrightarrow B_{p \cap B}$, we may assume that B is local with maximal ideal m and $p = (m, X)R$. In this case, one has

$$\begin{aligned}
\ell(I_{(m,X)R}) &= \dim \mathrm{gr}_{I_{(m,X)R}}(R_{(m,X)R}) \otimes R/(m,X)R \\
&= \dim \mathrm{gr}_I(R) \otimes R/(m,X)R \\
&\leq \dim \mathrm{gr}_I(R) \otimes R/XR \\
&\leq \dim B + \mathrm{rk}\,\Omega.
\end{aligned}$$

Here Ω stands for the underlying poset ideal that generates I. The last inequality comes from [BV, (5.10)] and from typical arguments based on the fact that Ω is a *straightening-closed* poset ideal (cf. [BST, (3.3.3)] for the technical details). On the other hand, $\mathrm{rk}\,\Omega = \mathrm{ht}(I) = (m-s+1)(n-s+1) - (m-s)(n-s)$, while $\mathrm{ht}(m,X)R = \dim B + mn - (m-s)(n-s)$. Therefore, we are to show that the difference $mn - (m - s + 1)(n - s + 1) = (m + n - s + 1)(s - 1)$ is at least two. Clearly, this is always the case for $s > 1$.

Assume now that $p \not\supset X$. Then, by the well-known procedure of inverting a variable and effecting elementary transformations [BV, Proposition (2.4)], one is reduced to a similar situation only with s decreased by one, whence the inductive hypothesis applies.

Therefore, one is left with the case where $s = 1$. But here, there is a natural isomorphism $\mathrm{gr}_I(R) \simeq R$ - this is true, more generally, for any graded homogeneous algebra and its irrelevant ideal. A similar reduction as above allows us to assume that $p = (p \cap B, X)R$ and, in this case, a computation shows that $\ell(I_p) = \mathrm{ht}(I)$, which proves our contention.

(ii) This part follows immediately from Proposition (2.6) and from the calculation of divisor class groups of determinantal rings [BV, (8.4)]. ∎

6.2. Slim ideals of maximal minors.

By a *slim* ideals of maximal minors is meant the ideal of the determinantal ring $R := B[X]/I_{s+1}(X)$ generated by the $s \times s$ minors of the first s rows of X. Recall that this ideal determines completely the canonical module of R [BV, (8.8)]; in particular, it is an ideal of height one. We of course make the proviso that $1 \leq s < min\{m, n\}$.

(6.2.1) Proposition. *Let $I \subset R$ be a slim ideal of maximal minors. Assume that the base ring B is a Cohen-Macaulay normal domain. Then:*
(i) $\mathrm{gr}_I(R)$ is a normal domain.
(ii) $\mathcal{R}(I)$ is normal and $\mathrm{Cl}(\mathcal{R}(I)) \simeq \mathrm{Cl}(B) \oplus \mathbb{Z}$.

PROOF: (i) The procedure is similar to the one employed in the proof of the preceding result: since $R/I \simeq B[X]/(I_s(X') + I_{s+1}(X))$ is again a normal domain [HE], where X' is the submatrix of X consisting of the first s rows, and $\mathrm{gr}_I(R)$ is Cohen-Macaulay, it is enough to check the earlier bounds for the local analytic spreads.

We induct on s. For $s > 1$, as long as the prime ideal $p \supset I$ in question does not contain $I_1(X')$, one can apply the inversion and elementary transformation trick. Therefore, assume that $p \supset I_1(X')$. We may assume as well that p is homogeneous and that B is local with maximal ideal $m = p \cap B$. Then $\mathrm{ht}(p) \geq \mathrm{ht}(m, I_1(X'))R = \dim B + \mathrm{ht}(I_1(X')R) \geq \dim B + 3$, the latter inequality being comfortably valid by [HE]. On the other hand, as in the previous proof, $l(I_p) \leq \ell(I_{(m,X)}) \leq \dim B + \mathrm{ht}(I) = \dim B + 1$. Thus, $\ell(I_p) \leq \mathrm{ht}(p) - 2$ in this case, as required.

One is left with the case $s = 1$. Here I is generated by the elements of the first row of X and any 2×2 minor involving this row gives a relation for two generators of I. Therefore, if the prime $p \supset I$ does not contain $I_1(X)$, I_p is principal. So, assume $p \supset I_1(X)$. But then, once more, $\mathrm{ht}(p) \geq \dim B + \mathrm{ht}(I_1(X)) - \mathrm{ht}(I_2(X)) = \dim B + m + n - 1 \geq \dim B + 3$. Therefore, $\ell(I_p) \leq \max\{\mathrm{ht}(I), \mathrm{ht}(p) - 2\}$ in all cases.

(ii) The normality of $\mathcal{R}(I)$ follow as before. The expression for the divisor class group of $\mathcal{R}(I)$ follows immediately from Proposition (2.6) and the previously cited result of [BV]. ∎

(6.2.2) Remark. The calculation of the divisor class group of $\mathrm{gr}_I(R)$ for both the slim and the full cases is a lot more involved [BST, (4.1.1)].

References

[BST] W. BRUNS, A. SIMIS AND N. V. TRUNG, *Blow-up of straightening-closed ideals in ordinal Hodge algebras*, Trans. Amer. Math. Soc. (to appear).

[BV] W. BRUNS AND U. VETTER, "Determinantal rings," Lecture Notes in Mathematics **1327** (Subseries: IMPA, Rio de Janeiro), Springer, Berlin-Heidelberg-New York, 1988.

[EH] D. EISENBUD AND C. HUNEKE, *Cohen-Macaulay Rees algebras and their specializations*, J. Algebra **81** (1983), 202–224.

[Har] F. HARARY, "Graph theory," Addison-Wesley Series in Mathematics, Addison-Wesley, Reading, Mass., 1972.

[HE] M. HOCHSTER AND J. EAGON, *Cohen-Macaulay rings, invariant theory and the generic perfection of determinantal loci*, Amer. J. Math. **93** (1971), 1020–1058.

[HSV 1] J. HERZOG, A. SIMIS AND W. V. VASCONCELOS, *Approximation complexes of blowing-up rings*, J. Algebra **74** (1982), 466–493.

[HSV 2] ——————, *Koszul homology and blowing-up rings*, in "Proceedings of Trento: Commutative Algebra," Lectures Notes in Pure and Applied Mathematics, Marcel-Dekker, New York, 1983, pp. 79–169.

[HSV 3] ——————, *On the arithmetic and homology of algebras of linear type*, Trans. Amer. Math. Soc. **283** (1984), 661–683.

[HSV 4] ——————, *Arithmetic of normal Rees algebras*.

[Hu 1] C. HUNEKE, *Powers of ideals generated by weak d-sequences*, J. Algebra **68** (1981), 471–509.

[Hu 2] ————, *Determinantal ideals of linear type*, Arch. Math. **47** (1986), 324–329.

[Hu 3] ————, *Hilbert functions and symbolic powers*, Michigan Math. J. **34** (1987), 293-318.

[HuRo] C. HUNEKE AND M. E. ROSSI, *The dimension and components of symmetric algebras*, J. Algebra **98** (1986), 200–210.

[HuS] C. HUNEKE AND J. SALLY, *Birational extensions in dimension two and integrally closed ideals*, J. Algebra **115** (1988), 491-500.

[HuSV] C. HUNEKE, A. SIMIS AND W. VASCONCELOS, *Reduced normal cones are domains*, in "Proceedings of an AMS Special Session: Invariant Theory (R. Fossum, M. Hochster and V. Lakshmibai, Eds.)," Contemporary Mathematics, American Mathematical Society, Providence, RI, 1989.

[KM] G. KEMPF, D. MUMFORD, ET AL., "Toroidal embeddings I," Lecture Notes in Mathematics, Springer-Verlag, Berlin, 1973.

[LT] M. LEJEUNE AND B. TEISSIER, "Clôture intégrale d'idéaux et équisingularité," Séminaire au Centre de Mathématiques de l'Ecole Politechnique, 1974.

[Mor] M. MORALES, *Noetherian symbolic blow-ups and examples in any dimension*, Max-Planck-Institut Preprint Series.

[Si] A. SIMIS, "Selected topics in Commutative Algebra," Lecture Notes, IX ELAM, Santiago, Chile, 1988.

[ST] A. SIMIS AND N. V. TRUNG, *The divisor class group of ordinary and symbolic blow-ups*, Math. Zeit. **198** (1988), 479–491.

[SV] A. SIMIS AND W. VASCONCELOS, *The Krull dimension and integrality of symmetric algebras*, Manuscripta Math. **61** (1988), 63–78.

[Va] W. VASCONCELOS, *Symmetric algebras*, in "This volume."

[Vi] R. VILLARREAL, *Cohen-Macaulay graphs*, Preprint.

Keywords. associated graded ring, catalectican matrix, Cohen-Macaulay, divisor class group, generic matrix, Gorenstein, graph, Hodge algebra, maximal minor, monomial, normal, rank, Rees algebra, straightening-closed ideal.
1980 *Mathematics subject classifications*: 13C05, 13C13, 13C15, 13H10

Universidade Federal da Bahia, Instituto de Matemática, Av. Ademar de Barros, s/n, 40210 Salvador, Bahia, Brazil

Symmetric Algebras

WOLMER V. VASCONCELOS*

Contents

1   Krull dimension 116
    1.1   The Forster–Swan number 116
    1.2   Ideals of linear type 119
    1.3   Dimension formulas 121
2   Integral domains 123
    2.1   Irreducibility 123
    2.2   Commuting varieties 124
    2.3   Complete intersections 126
    2.4   Almost complete intersections 127
    2.5   Modules of projective dimension two 129
    2.6   Approximation complexes 133
3   Jacobian criteria 135
    3.1   Regular primes 136
    3.2   Normality 138
    3.3   Divisor class group 139
4   Factoriality 141
    4.1   The factorial conjecture 141
    4.2   Homological rigidity 143
    4.3   Finiteness of ideal transforms 146
    4.4   Symbolic power algebras 153
    4.5   Roberts construction 155

Introduction

Given a commutative ring R and an R–module E, the symmetric algebra of E is an R–algebra $S(E)$ together with a R–module homomorphism $\pi : E \longrightarrow S(E)$ that solves the following universal problem. For a commutative R-algebra B and any R–module homomorphism $\varphi : E \longrightarrow B$, there exists a unique R–algebra homomorphism $\Phi : S(E) \longrightarrow B$ such that the diagram

$$
\begin{array}{ccc}
E & \xrightarrow{\varphi} & B \\
\pi \downarrow & \nearrow \Phi & \\
S(E) & &
\end{array}
$$

is commutative.

Thus, if E is a free module, $S(E)$ is a polynomial ring $R[T_1, \ldots, T_n]$, one variable for each element in a given basis of E. More generally, when E is given by the presentation

---

*This research was partially supported by the National Science Foundation.

$$R^m \xrightarrow{\varphi} R^n \longrightarrow E \longrightarrow 0, \ \varphi = (a_{ij}),$$

its symmetric algebra is the quotient of the polynomial ring $R[T_1, \cdots, T_n]$ by the ideal $J(E)$ generated by the 1-forms

$$f_j = a_{1j}T_1 + \cdots + a_{nj}T_n, \ j = 1, \ldots, m.$$

Conversely, any quotient ring of a polynomial ring $R[T_1, \ldots, T_n]/J$, with J generated by 1-forms in the T_i's, is the symmetric algebra of a module.

These algebras began to be systematically studied in Micali's thesis [41] (see also [42] and [4]), whose most pertinent result for us is the characterization of smooth closed points of an affine variety by the integrality of the symmetric algebra of its ideal of definition.

Much later, Huneke [28] and Valla [53] showed that the integrality of symmetric algebras of ideals was strongly connected with a far–reaching generalization of the notion of regular sequence—*d-sequences*—that shared most of the analytic properties of the former, and occurred in greater profusion. Two other developments were the introduction of the *approximation complexes* ([50], [19])—a family of differential graded complexes that attempts to measure the difference between the symmetric and Rees algebras of an ideal—and, the role of *linkage* in the work of Huneke, Ulrich, and others, that served to unify several of the aspects of the area. An early review of these developments is [20].

The emphasis here is on modules of rank at least two; symmetric algebras of ideals will only be mentioned when comparisons are unavoidable. This results in a diminished role for Koszul homology and linkage, but in return certain features not usually present, *e.g.* reflexivity of modules, will come to the fore.

The running theme proper is the ideal theory—Krull dimension, integrality, normality and factorization—of symmetric algebras, and some modifications of them. It highlights some of known results, oftentimes their proofs, and lists significant open problems.

Individual sections having their own introductions, we limit ourselves here to a sketchy picture of its contents. Section 1 discusses the Krull dimension of a symmetric algebra $S(E)$. It is fairly complete in that there is an abstract general formula, and, based on it, a constructive method for computing the dimension in terms of a presentation of the module. From then on the gaps abound, beginning with a rudimentary theory of when an ideal generated by the 1–forms above is prime. We attempt to remedy this state of affairs by sketching the elements of the theory of the approximation complexes, with examples of its applications. But already for modules of projective dimension two there is no general method to ascertain when $J(E)$ is prime. This is unfortunate because, as will be pointed out, there are many interesting situations in need of resolution. Section 3 discusses Jacobian criteria, with applications to normality and the determination of the divisor class group of normal algebras. An application is the proof of the *Zariski–Lipman conjecture* for symmetric algebras over polynomial rings. The next section discusses the factoriality of $S(E)$ and the finiteness question of the factorial closure of a symmetric algebra. An early finiteness conjecture was settled in the negative by a especially beautiful example of Roberts [47]. It manages to provide counterexamples to questions ranging from Hilbert's 14th Problem to the finiteness of symbolic blow–ups. We describe some of its fine details. A great deal of emphasis is also placed on explicit methods to compute ideal transforms.

Several colleagues have made useful comments on earlier drafts of this survey. I am indebted to Jürgen Herzog, Aron Simis and Rafael Villarreal; as fellow co-conspirators on these matters, they cannot be fully excused of its shortcomings.

1   Krull dimension

The Krull dimension of a symmetric algebra of a module E is connected to an invariant $b(E)$ introduced by Forster [14] a quarter of century ago, that bounds its number of generators. This was done recently by Huneke and Rossi [31]; furthermore, it was accomplished in a manner that makes the search for dimension formulas for $S(E)$ much easier.

Based on slightly different ideas [52] gives another proof of that result, and while in [51] one only estimated the Krull dimension of $S(E)$ in terms of the heights of the Fitting ideals of E, [52, Theorem 1.1.4] gives the exact formula. It makes for an often effective way of determining the Krull dimension of $S(E)$.

1.1   The Forster–Swan number

To be more precise, let be given a Noetherian ring R of finite Krull dimension, and let E be a finitely generated R-module.

For a prime ideal P of R, denote by $v(E_P)$ the minimal number of generators of the localization of E at P. It is the same as the torsion free rank of the module E/PE over the ring R/P. The Forster–Swan number of E is:

$$b(E) = \sup_{P \in Spec(R)} \{\dim(R/P) + v(E_P)\}.$$

The original result of Forster was that E can be globally generated by $b(E)$ elements. Later it was refined, at the hands of Swan and others, by appropriately restricting the set of primes. It has to this day played a role in algebraic K-theory. The theorem of Huneke–Rossi [31] explains the nature of this number.

Theorem 1.1.1 $\dim S(E) = b(E)$.

To give the proof of this we need a few preliminary observations about dimension formulas for graded rings. M. Kühl has also noticed these formulas.

Lemma 1.1.2 *Let B be a Noetherian integral domain that is finitely generated over a subring A. Suppose there exists a prime ideal Q of B such that $B = A + Q$, $A \cap Q = 0$. Then*

$$\dim(B) = \dim(A) + \text{height}(Q) = \dim(A) + \text{tr.deg.}_A(B).$$

Proof. We may assume that $\dim A$ is finite; $\dim(B) \geq \dim(A) + \text{height}(Q)$ by our assumption. On the other hand, by the standard dimension formula of [40], for any prime ideal P of B, $\mathbf{p} = P \cap A$, we have

$$\text{height}(P) \leq \text{height}(\mathbf{p}) + \text{tr.deg.}_A(B) - \text{tr.deg.}_{k(\mathbf{p})} k(P).$$

The inequality $\dim(B) \leq \dim(A) + \operatorname{height}(Q)$ follows from this formula and reduction to the affine algebra obtained by localizing B at the zero ideal of A. $\square$

There are two cases of interest here. If B is a Noetherian graded ring and A denotes its degree 0 component then:

$$\dim(B/P) = \dim(A/\mathbf{p}) + \operatorname{tr.deg.}_{k(\mathbf{p})} k(P)$$

and

$$\dim(B_{\mathbf{p}}) = \dim(A_{\mathbf{p}}) + \operatorname{tr.deg.}_A(B).$$

We shall need to identify the prime ideals of $S(E)$ that correspond to the extended primes in case E is a free module. It is based on the observation that if R is an integral domain, then the R–torsion submodule T of a symmetric algebra $S(E)$ is a prime ideal of $S(E)$. This is clear from the embedding $S(E)/T \hookrightarrow S(E) \otimes K$ ($K = $ field of quotients of R) and the latter being a polynomial ring over K.

Let $\mathbf{p}$ be a prime ideal of R; denote by $T(\mathbf{p})$ the $R/\mathbf{p}$–torsion submodule of

$$S(E) \otimes R/\mathbf{p} = S_{R/\mathbf{p}}(E/\mathbf{p}E).$$

The torsion submodule of $S(E)$ is just $T(0)$.

Proof of theorem. By the formula above we have

$$
\begin{aligned}
\dim S(E)/T(\mathbf{p}) &= \dim(R/\mathbf{p}) + \operatorname{tr.deg.}_{R/\mathbf{p}} S(E)/T(\mathbf{p}) \\
&= \dim(R/\mathbf{p}) + v(E_{\mathbf{p}})
\end{aligned}
$$

and it follows that $\dim S(E) \geq b(E)$.

Conversely, let P be a prime of $S(E)$ and put $\mathbf{p} = P \bigcap R$. It is clear that $T(\mathbf{p}) \subset P$,

$$\dim S(E)/P \leq \dim S(E)/T(\mathbf{p}),$$

and $\dim S(E) \leq b(E)$ as desired. $\square$

Corollary 1.1.3 *Let R be a local domain of dimension d, and let E be a finitely generated module that is free on the punctured spectrum of R. Then $\dim S(E) = \sup\{v(E), d + \operatorname{rank}(E)\}$.*

Another way to derive the dimension of $S(E)$ is through the following valuation theoretic argument (see [17] for a treatment of valuative dimensions). The Krull dimension of the Noetherian ring $S(E)/T(P)$ is the supremum of the Krull dimension of all of its valuation rings. If A is one such and V is its restriction to a valuation of R/P, then the subring of A generated by V and the degree one component M of $S(E)/T(P)$ is a polynomial ring over V on as many variables as the rank of M over R/P, since VM is a free V–module.

Let us give an application due to R. Villarreal [61] of this formula. Given a graph $\mathcal{G}$ on a vertex set $V = \{X_1, \ldots, X_n\}$, he attaches the ideal $I(\mathcal{G})$ of the polynomial ring $k[X_1, \ldots, X_n]$, generated by the monomials $X_i X_j$ defined by its edges. This is not the same as the Reisner–Stanley ideal of the simplicial complex defined by $\mathcal{G}$. Instead, it is related to its complementary simplex $\mathcal{G}'$: if the latter has no triangles, then the two ideals coincide.

Theorem 1.1.4 *Let $\mathcal{G}$ be a graph with n vertices and q edges, and let $I = I(\mathcal{G})$. If $\mathcal{G}$ is connected, then the Krull dimension of $S(I)$ is $sup\{n+1, q\}$.*

Proof. It is clear that the Krull dimension of $S(I)$ is at least the given bound. To prove the converse, we recall the notion of a minimal vertex cover of a graph. It is simply a subset $\mathcal{A}$ of $\mathcal{G}$ such that every edge of $\mathcal{G}$ is incident with a vertex of $\mathcal{A}$ and admits no proper subset with this property. There is a one–to–one correspondence between the minimal covers of $\mathcal{G}$ and the minimal primes of $I(\mathcal{G})$.

Let P be a prime ideal of height $n - i$, containing I; set $\mathcal{B} = \{v \in V \mid v \notin P\}$. Let $Q = (\mathcal{A})$ be a minimal prime of I contained in P, where $\mathcal{A}$ is a minimal vertex cover of $\mathcal{G}$. Define now

$$\mathcal{C} = \{x \in \mathcal{A} \mid x \text{ is adjacent to some vertex in } \mathcal{B}\}$$

and

$$Y = \{\{x, y\} \in X(\mathcal{G}) \mid \{x, y\} \bigcap \mathcal{C} = \emptyset.$$

Notice that $v(I_P) \leq |\mathcal{C}| + |Y|$. Since $|\mathcal{B}| \geq i$, we obtain

$$v(I_P) + \dim(R/P) \leq |\mathcal{C}| + |Y| + |\mathcal{B}|.$$

The formula is now a consequence of the following.

Proposition 1.1.5 *Let $\mathcal{G}$ be a connected (n, q)-graph with vertex set $V = V(\mathcal{G})$ and edge set $X = X(\mathcal{G})$. Let $\mathcal{A}$ be a given minimal vertex cover of $\mathcal{G}$, and let $\mathcal{B}$ be a subset of $V \setminus \mathcal{A}$. Then $|\mathcal{B}| + |\mathcal{C}| + |Y| \leq sup\{n+1, q\}$.*

Proof. Let Y' be the set of edges covered by $\mathcal{C}$, that is, $Y' = X(\mathcal{G}) \setminus Y$. Consider the subgraph $\mathcal{G}'$ of $\mathcal{G}$ with edge set equal to Y' and vertex set

$$V(\mathcal{G}') = \{z \in V(\mathcal{G}) \mid z \text{ lies in some edge in } Y'\}.$$

If $Y = \emptyset$, then $|\mathcal{B}| + |\mathcal{C}| + |Y| \leq n$. Assume $Y \neq \emptyset$, and denote the connected components of Y' by $\mathcal{G}_1, \ldots, \mathcal{G}_m$. Set $\mathcal{B}_i = \mathcal{B} \bigcap V(\mathcal{G}_i)$ and $\mathcal{C}_i = \mathcal{C} \bigcap V(\mathcal{G}_i)$; notice $\mathcal{B} = \bigcup \mathcal{B}_i$ and $\mathcal{C} = \bigcup \mathcal{C}_i$.

We claim that $|\mathcal{B}_i| + |\mathcal{C}_i| \leq n_i - 1$, for all i, where $n_i = |V(\mathcal{G}_i)|$. For that, fix an edge $\{x, y\} \in Y$. If $x \in V(\mathcal{G}_i)$, $x \notin \mathcal{B}_i \bigcup \mathcal{C}_i$; using that $\mathcal{B}_i$ and $\mathcal{C}_i$ are disjoint we get the asserted inequality. On the other hand, if $x \notin V(\mathcal{G}_i)$, we choose a vertex $z \in V(\mathcal{G}_i)$ at a minimum distance from x. This yields a path $\{x = x_0, x_1, \cdots, x_r = z\}$. As $x_{r-1} \notin V(\mathcal{G}_i)$, $z \notin \mathcal{B}_i \bigcup \mathcal{C}_i$, which gives $|\mathcal{B}_i| + |\mathcal{C}_i| \leq n_i - 1$. Altogether we have

$$|\mathcal{B}| + |\mathcal{C}| + |Y| \leq \sum_{i=1}^{m}(n_i - 1) + |Y| = \sum_{i=1}^{m}(n_i - 1) + (q - |Y'|).$$

But Y' is the disjoint union of $X(\mathcal{G}_i)$ for $i = 1, \ldots, m$, and thus $|Y'| = \sum_{i=1}^{m} q_i$, $q_i = |X(\mathcal{G}_i)|$. This permits writing the last inequality as

$$|\mathcal{B}| + |\mathcal{C}| + |Y| \leq \sum_{i=1}^{m}(n_i - 1) + q - \sum_{i=1}^{m} q_i = \sum_{i=1}^{m}(n_i - q_i - 1) + q.$$

Since $\mathcal{G}_i$ is a connected (n_i, q_i)-graph we have $n_i - 1 \leq q_i$. Therefore $|\mathcal{B}| + |\mathcal{C}| + |Y| \leq q$, establishing the claim. $\square$

1.2 Ideals of linear type

We shall now connect Theorem 1.1.1 to an abstract dimension formula. First a definition is recalled. For an ideal I of a commutative ring R there exists a canonical mapping

$$\alpha : S(I) \longrightarrow \mathcal{R}(I)$$

from the symmetric algebra of I onto its Rees algebra. I is said to be of *linear type* if α is an isomorphism. This terminology was originally introduced by G. Valla and L. Robbiano.

Theorem 1.2.1 ([52, Theorem 1.2.1]) *Let R be a Noetherian ring and let I be an ideal of linear type contained in the Jacobson radical of R. Then*

$$\dim(R) = \sup_{P \supseteq I} \{\dim(R/P) + v(I_P)\}.$$

Proof. Since I is of linear type, one has $S_{R/I}(I/I^2) = gr_I(R)$. On the other hand, $\dim gr_I(R) = \dim R$, cf. [40, page 122]. Applying Theorem 1.1.1 one obtains the assertion. $\square$

As it will be seen, Theorem 1.1.1 is, in turn, a consequence of this formula. The interplay provided by this equivalence is rewarding; for instance, it provides for a very short proof of Theorem 1.1.1 when R is catenarian.

Given an R-module E, the irrelevant ideal $I = S(E)_+$ of the symmetric algebra $B = S(E)$ is of linear type, cf. [20, page 87]. Note

$$S(E) = S_{B/I}(I/I^2)$$

so that if we localize B at the multiplicative set $1 + I$ we ensure all the hypotheses of the theorem above. What remains is to observe that each prime ideal of $S(E)$ that contains $S(E)_+$ has the form $\mathbf{p} + S(E)_+$, where $\mathbf{p}$ is a prime ideal of R.

The earliest significant example of an ideal of linear type is found in [41]: Every ideal generated by a regular sequence has this property (see [21] for further details). This result had a far-reaching generalization.

Definition 1.2.2 Suppose $\mathbf{x} = \{x_1, \ldots, x_n\}$ is a sequence of elements in a ring R. The sequence $\mathbf{x}$ is called a *d–sequence* if

1. $\mathbf{x}$ is a minimal generating system of the ideal $(x_1, \ldots, x_n)$.

2. $(x_1, \ldots, x_i) : x_{i+1}x_k = (x_1, \ldots, x_i) : x_k$ for $i = 0, \ldots, n-1$ and $k \geq i + 1$.

Theorem 1.2.3 ([28], [53]) *Every ideal generated by a d–sequence is of linear type.*

The similarity between *d–sequences* and *regular sequences* is further enhanced when the former are interpreted in terms of the vanishing of certain complexes (see Theorem 2.6.2) derived from Koszul complexes.

There are many other known classes of ideals of linear type. The determinantal ideals associated to a generic matrix which are of linear type have been fully described in [27].

More recently, B. Kotsev has proved that the ideal generated by the submaximal minors of a generic symmetric matrix is of linear type (over arbitrary base rings) [35]. For ideals attached to graphs, there is an emerging theory, see [61].

The generic determinantal ideals are also noteworthy because they cannot be generated by *d–sequences*. It is not known what happens in the symmetric case—that is, whether they can be generated by a *d–sequence*.

The definition of linear type can be refined as follows. Consider the canonical mapping from $S(I)$ onto $R(I)$:

$$0 \longrightarrow \mathcal{A} \longrightarrow S(I) \longrightarrow R(I) \longrightarrow 0.$$

The kernel $\mathcal{A}$ is a graded ideal

$$\mathcal{A} = \sum_{i \geq 2} \mathcal{A}_i,$$

with trivial component in degree 1. Very often the degree 2 component vanishes as well—and the ideal will be called *syzygetic*.

Definition 1.2.4 I is said to be of rth *type*, if $\mathcal{A}$ is generated by its components of degree at most r.

Are there interesting cases of ideals of higher type? There is some fragmentary information for ideals of *cubic* type.

1.3 Dimension formulas

Let R be a Noetherian domain and let

$$\varphi : R^m \longrightarrow R^n$$

be a presentation of the R-module E. We intend to express $\dim S(E)$ in terms of the sizes of the determinantal ideals of the matrix φ.

There is a set of conditions on the sizes of the Fitting ideals of E that keep recurring. For each integer $t \geq 1$ denote by $I_t(\varphi)$ the ideal generated by the $t \times t$ minors of φ. For some non-negative integer k, consider the condition $\mathcal{F}_k$:

$$\text{height}(I_t(\varphi)) \geq \text{rank}(\varphi) - t + 1 + k, \quad 1 \leq t \leq \text{rank}(\varphi).$$

We recall that the classical bound for the sizes of these ideals is given by the theorem of Eagon and Northcott ([11]): $\text{height}(I_t(\varphi)) \leq (m - t + 1)(n - t + 1)$, with equality reached when φ is a generic matrix in $m \cdot n$ indeterminates.

One can rephrase $\mathcal{F}_k$ in terms of how L, the image of φ, embeds into R^n. It means that for each prime ideal P where the localization E_P is not a free module then

$$\nu(E_P) \leq \text{rank}(E) + \text{height}(P) - k.$$

This says that for each prime ideal P of R, L_P decomposes into a summand of R_P^n and a submodule K of rank at most $\text{height}(P) - k$ (*cf.* [20], [52]). In particular, if E is an ideal containing a regular element, then it cannot satisfy $\mathcal{F}_2$, as otherwise Krull's principal ideal theorem would be violated.

Example 1.3.1 W. Bruns pointed out the following procedure to obtain examples of torsion–free modules with $\mathcal{F}_k$ for various values of k.

Let R be a local domain of dimension d, and let G be the module of global sections of a vector bundle on the punctured spectrum of R, of rank $d - k$. From a presentation

$$0 \longrightarrow L \longrightarrow R^n \overset{\varphi}{\longrightarrow} G \longrightarrow 0$$

of G, the module $E = coker(\varphi^*)$, by the remark above, satisfies $\mathcal{F}_k$.

Applied to the module of Example 2.4.5, one gets a torsion–free module (even reflexive in this case) with the condition $\mathcal{F}_2$. On the other hand, when applied to the module of the Horrocks–Mumford bundle, one obtains a module with $\mathcal{F}_3$.

We now return to the derivation of a dimension formula for $S(E)$. A lower bound for $b(E)$ is $b_0(E) = \dim R + \text{rank}(E)$, the value corresponding to the generic prime ideal in the definition of $b(E)$. We show that the correction from $b_0(E)$ can be explained by how deeply the condition $\mathcal{F}_0$ is violated. Set $m_0 = \text{rank}(\varphi)$, so that $\text{rank}(E) = n - m_0$. Without loss of generality we assume that R is a local ring, $\dim R = d$. Consider the descending chain of affine closed sets:

$$V(I_{m_0}(\varphi)) \supseteq \cdots \supseteq V(I_1(\varphi)) \supseteq V(1).$$

Let $P \in Spec(R)$; if $I_{m_0}(\varphi) \not\subseteq P$, $\text{rank}(E/PE) = n - m_0$, and therefore

$$\dim(R/P) + \text{rank}(E/PE) \le b_0(E).$$

On the other hand, if $P \in V(I_t(\varphi)) \backslash V(I_{t-1}(\varphi))$, we have $\text{rank}(E/PE) = n - t + 1$; if $\mathcal{F}_0$ holds at t, the height of P is at least $m_0 - t + 1$ and again $\dim(R/P) + \text{rank}(E/PE) \le b_0(E)$.

Define the following integer valued function on $[1, \text{rank}(\varphi)]$:

$$d(t) = \begin{cases} m_0 - t + 1 - \text{height}(I_t(\varphi)) & \text{if } \mathcal{F}_0 \text{ is violated at } t \\ 0 & \text{otherwise.} \end{cases}$$

Finally, if we put $d(E) = \sup_t\{d(t)\}$, we have the following dimension formula.

Theorem 1.3.2 ([52, Theorem 1.1.2]) *Let R be an equi–dimensional catenarian domain and let E be a finitely generated R-module. Then*

$$b(E) = b_0(E) + d(E).$$

Proof. Assume that $\mathcal{F}_0$ fails at t, and let P be a prime as above. We have

$$\text{height}(P) \ge m_0 - t + 1 - d(t),$$

and thus

$$\dim(R/P) + \text{rank}(E/PE) \le (d - (m_0 - t + 1 - d(t))) + (n - t + 1) = b_0(E) + d(t).$$

Conversely, pick t to be an integer where the *largest* deficit $d(t)$ occurs. Let P be a prime ideal minimal over $I_t(\varphi)$ of height exactly $m_0 - t + 1 - d(t)$. From the choice of t it follows that $P \notin V(I_{t-1}(\varphi))$, as otherwise the deficit at $t - 1$ would be even higher. Since R is catenarian the last displayed expression gives the desired equality. $\square$

Remark 1.3.3 If R is not an integral domain, let $p_1, \ldots, p_n$ be its minimal primes. It is clear from Theorem 1.1.1 that

$$\dim S_R(E) = \sup_{i=1}^{n}\{\dim S_{R/p_i}(E/p_iE)\}.$$

Assume that for each p_i, R/P_i is equi-dimensional. If we put

$$b_i(E) = b_0(E/p_iE), \quad d_i(E) = d(E/p_iE),$$

then

$$\dim S(E) = \sup_{i=1}^{n}\{b_i(E) + d_i(E)\}.$$

Example 1.3.4 Here is a simple illustration: If $E = \operatorname{coker}(\varphi)$

$$\varphi = \begin{bmatrix} 0 & x & x \\ y & z & 0 \\ x & 0 & x \end{bmatrix},$$

$m_0 = 3, d(1) = d(3) = 0$, *but* $d(2) = 1$, so $\dim S(E) = 3 + 1$.

One can define the condition $\mathcal{F}_k$ for negative integers as well (*cf.* [49]). The number $d(E)$, if positive, determines the most strict of the conditions satisfied by E: If $k = -d(E)$, $\mathcal{F}_k$ holds but $\mathcal{F}_{k+1}$ does not.

Remark 1.3.5 We can express $\mathcal{F}_k$ in terms of Krull dimension: If E has $\mathcal{F}_k$ then for any ideal I of R, of height at least k, height $(IS(E)) \geq k$.

We shall now observe that the formula fails for non-catenarian domains. Pick a local domain of dimension three with two saturated chains of primes as in the graph (see [44, Example 2]):

Let E be the module $R/P \oplus R/P \oplus R/Q$. A simple calculation shows that $b(E) = b_0(E) = 3$. If we present it as

$$R^m \xrightarrow{\varphi} R^3,$$

$I_2(\varphi) = P(P+Q)$ has height 1, so that $d(1) = 3 - 2 + 1 - 1 = 1$.

2 Integral domains

The conditions under which the symmetric algebra $S(E)$ is an integral domain have been a source of interest. The situation is well understood for modules of projective dimension one (*cf.* [3], [26], [51]), for several classes of ideals (see [20] for a survey) and have been an incentive to the development of several generalizations of the notion of regular sequences and the theory of the approximation complexes.

Throughout this section R will be a Cohen-Macaulay integral domain, although the full strength of this condition will not always be used.

2.1 Irreducibility

We have found the search for a general condition on E that leads $S(E)$ to be an integral domain to be complicated by the many diverse ways it occurs. Nevertheless, the following result (*cf.* [6, Proposition 2.2]) identifies a basic ingredient.

Proposition 2.1.1 *Let R be a catenarian domain whose maximal ideals have the same height and let E be a finitely generated R-module. Then $Spec(S(E))$ is irreducible if and only if E satisfies $\mathcal{F}_1$ and all the minimal primes of $S(E)$ have the same dimension.*

Proof. One of the minimal prime ideals of a symmetric algebra $S(E)$ is the R–torsion submodule T of $S(E)$. Since the Krull dimension of $S(E)/T$ is $\dim R +$ rank E (*cf.* [31], [52]), it follows that if $S(E)$ is equi–dimensional then by Theorem 1.3.2 the condition $\mathcal{F}_0$ is automatically satisfied. Thus both conditions in the assertion imply $\mathcal{F}_0$.

Suppose $Spec(S(E))$ is irreducible; then, for each nonzero element x of R, as $x \notin T$, $\dim S(E) \otimes R/(x) = \dim S(E) - 1$. But $S_R(E) \otimes R/(x) \cong S_{R/(x)}(E/xE)$, so that the algebra $S(E/xE)$ will satisfy the condition of Theorem 1.3.2 if $R/(x)$ is reduced. It is not difficult to see that we can pick a square-free element x contained in all the associated primes of the $I_t(\varphi)$'s. The condition $\mathcal{F}_1$ will follow.

Conversely, suppose $\mathcal{F}_1$ holds and M is a minimal prime of $S(E)$ other than T. If $P = M \bigcap R$, then M is just $T(P)$. If however $P \neq 0$, reducing E modulo a prime element x of P would, in the presence of $\mathcal{F}_1$, yield an $R/(x)$–module E/xE whose $R/(x)$–rank is still rank E, so that the dimension of $S(E/xE)$, by Theorem 1.3.2, is one less than that of $S(E)$. The equi–dimensionality hypothesis on $S(E)$ rules this out. $\square$

Corollary 2.1.2 *Let R be an integral domain. If E satisfies $\mathcal{F}_1$ and $S(E)$ is Cohen–Macaulay, then $S(E)$ is an integral domain.*

Proof. At each localization of R $T(0)$ is the only minimal prime of $S(E)$ by the the result above. Since the assumption implies that it is the only associated prime as well, the assertion ensues. $\square$

2.2 Commuting varieties

A major class of examples of symmetric algebras is that associated to commuting varieties ([6]). They are defined as follows. Let V be a variety, endowed with an algebra structure, and let W be one of its linear subvarieties. Let $\{e_1, e_2, \ldots, e_n\}$ and $\{g_1, g_2, \ldots, g_m\}$ be bases of W and V, and consider two independent generic elements of W: $x = \sum x_i e_i$ and $y = \sum y_i e_i$.

The coordinates of their commutator

$$[x, y] = \sum_{i=1}^{i=m} f_i g_i$$

gives an ideal of definition

$$J(W) = (f_1, \ldots, f_m) \subset k[x_1, \ldots, x_n, y_1, \ldots, y_n]$$

for $C(W)$. We refer to $J(W)$ as the *obvious* equations for $C(W)$.

Denote by φ the Jacobian submatrix of $J(W)$ with respect to the the subset $\{y_1, y_2, \ldots, y_n\}$. It is a matrix of linear forms of the ring $R = k[x_1, x_2, \ldots, x_n]$.

Definition 2.2.1 The *Jacobian module* of the commuting variety of W is the R–module $E = \text{cokernel } (\varphi)$.

In other words, the ideal $J(W)$ can be represented in matrix notation

$$(f_1, \ldots, f_n) = (y_1, \ldots, y_n) \cdot \varphi,$$

where the entries of φ are the derivatives of the f_j's with respect to the y_i's. It is a matrix of 1–forms over the ring $R = k[x_1, \ldots, x_n]$. It defines a module E, the *Jacobian module of A*. (Actually, it is the dual of the usual Jacobian matrix.) Its significance lies in the following:

Proposition 2.2.2 ([6, Proposition 1.2]) *Let $S(E)$ denote the symmetric algebra of the R–module E. Then*

$$C(W) \simeq Spec(S(E))_{red}.$$

Whenever $W = V$, by abuse of terminology, we call E the Jacobian module of V. In this case there is another description of the Jacobian module. Let $\{e_1, \ldots, e_n\}$ be a basis of V, and denote by R the ring of regular functions on V, $R = k[x_1, \ldots, x_n]$. Put $x = \sum x_i e_i$. Dualizing the exact sequence

$$V \otimes R \xrightarrow{\varphi} V \otimes R \longrightarrow E \longrightarrow 0,$$

if we identify $V \otimes R$ and $(V \otimes R)^*$, it is easy to see that

$$\varphi^*(a) = adx(a) = [x, a],$$

x the generic element of above. Therefore the dual of E is the centralizer of the generic element of V

By way of illustration, let us consider

Example 2.2.3 (a) Let L be the 3-dimensional Lie algebra $\{e, f, g \}$ defined by

$$[ef] = 0, \quad [eg] = e, \quad [fg] = f.$$

$W = V$ is affine 3-space. The ideal $J(W)$ is defined by the forms $x_1 y_3 - x_3 y_1$ and $x_2 y_3 - x_3 y_2$. The Jacobian module E has a presentation

$$0 \longrightarrow R^2 \overset{\varphi}{\longrightarrow} R^3 \longrightarrow E \longrightarrow 0,$$

$$\varphi = \begin{pmatrix} -x_3 & 0 \\ 0 & -x_3 \\ x_1 & x_2 \end{pmatrix}.$$

It is easy to see that $S(E)$ is reduced, so that $\mathcal{C}(W) = Spec(S(E))$. It has two irreducible components.

(b) Denote by DS_n the space of all $n \times n$ matrices with equal line sums–that is, essentially doubly stochastic matrices. If $n = 3$, a calculation with the Bayer and Stillman *Macaulay* program ([5]) shows that the Jacobian module E of $\mathcal{C}(DS_3)$ has projective dimension two and $S(E)$ is a Cohen-Macaulay integral domain.

(c) Let V be an n-dimensional vector space over k. There is a natural Lie algebra structure on $L = V \oplus \wedge^2 V$ that makes $\wedge^2 V$ the center of L. The Jacobian module of L is the direct sum of a free module of rank $n(n-1)/2$, corresponding to its center, and the module that has n generators and for relations the forms $x_i y_j - x_j y_i$, $1 \le i < j \le n$—that is, the ideal $(x_1, \ldots, x_n)$. The projective dimension of Jacobian modules can thus attain any value.

(d) Finally, let H_n be the Heisenberg algebra of dimension $2n + 1$, with commutation relations $[P_i, Q_i] = E$, $i = 1, \ldots, n$. The commuting variety $\mathcal{C}(H_n)$ is defined by a single equation

$$\sum_{i=1}^{i=n} x_i y_{n+i} - x_{n+i} y_i.$$

It is a factorial variety for $n \ge 2$.

2.3   Complete intersections

For the symmetric algebras of modules of projective dimension 1, the main application of the complex above is:

Theorem 2.3.1 ([3], [26], [52]) *Let E be a module of projective dimension 1. The following conditions are equivalent.*

(a) $\mathcal{Z}(E)$ *is acyclic.*

(b) E *satisfies* $\mathcal{F}_0$.

(c) $S(E)$ *is a complete intersection.*

Moreover, if R is an integral domain, then $S(E)$ is an integral domain if and only if E satisfies the condition $\mathcal{F}_1$.

We now consider an example in detail (*cf.* [6]). Let k be an algebraically closed field, of characteristic 0, and let $S_n(k)$ be the affine space of all symmetric matrices of order n with entries in k. The commuting variety of $S_n(k)$ is defined by the ideal generated by the entries of

$$Z = [X, Y] = X \cdot Y - Y \cdot X,$$

where X and Y are generic symmetric $n \times n$ matrices in $n(n+1)$ indeterminates. $Z = [z_{ij}]$ is an alternating matrix of 2-forms.

Theorem 2.3.2 ([6, Theorem 3.1]) *The entries of Z form a regular sequence generating a prime ideal.*

The proof will follow from Theorem 2.3.1, after certain details of the structure of the Jacobian module are made clear.

Lemma 2.3.3 *The Jacobian module of $C(S_n(k))$ has projective dimension one.*

Proof. It suffices to show that the presentation matrix φ of E has rank $n(n-1)/2$. If we specialize the matrix X to a generic diagonal matrix, it is easy to see that the forms z_{ij} specialize to

$$z_{ij}^* = (x_{ii} - x_{jj})y_{ij},$$

so that the corresponding matrix has full rank. $\square$

Proposition 2.3.4 $C(S_n(k))$ *is an irreducible variety.*

Proof. It suffices to show that if W is a generic symmetric matrix commuting with X then the pair (X, W) is a generic point of $C(S_n(k))$. This is a formal consequence of the proof of [16, Theorem 1, p. 341-342] once Lemmas 2.3.5 and 2.3.6 have been established.

We recall that a square matrix is *non-derogatory* provided its minimal polynomial is its characteristic polynomial.

Lemma 2.3.5 *Let A be a square matrix. Then the following are equivalent.*

(a) *A is non-derogatory.*

(b) *If B is a matrix and $[A, B] = 0$, then there is a polynomial $p(t)$ with $p(A) = B$.*

Proof. This is [16, Proposition 4].

Lemma 2.3.6 *Let B be an element of $S_n(k)$. There exists a non-derogatory element of $S_n(k)$ that commutes with B.*

Proof. By [15, Corollary 2, p. 13], and the Jordan decomposition theorem, there exists an orthogonal matrix O such that

$$O^t BO = \oplus_{i=1}^s (\lambda_i I_i + N_i),$$

with N_i nilpotent, symmetric, and $\lambda_i I_i + N_i$ irreducible. (M^t is the transpose of the matrix M.) The matrix

$$O(\oplus_{i=1}^s (\mu_i I_i + N_i))O^t,$$

with distinct μ_i's, is non-derogatory and commutes with B. $\square$

For each integer $0 \le r \le n$ define

$$M_n^r = \{(A, B) \in S_n(k) \times S_n(k) \mid \text{rank } [A, B] \le r\}.$$

Corollary 2.3.7 ([6]) $M_n^{2r} = M_n^{2r+1}$ *is an irreducible Gorenstein variety of codimension* $(n - 2r - 1)(n - 2r)/2$. *Its reduced equations are the Pfaffians of Z of order $2r + 2$.*

Other examples of modules of projective dimension 1, derived from Lie algebras, are considered in [6].

Problem 1. The commuting variety of $S_n(k)$ is factorial for $n \le 5$. Is it so in general?

2.4   Almost complete intersections

For $S(E)$ to be an almost complete intersection requires that E be a module of projective dimension 2 whose second Betti number is 1. (This case was studied in [54].) :

$$0 \longrightarrow R \xrightarrow{\psi} R^m \xrightarrow{\varphi} R^n \longrightarrow E \longrightarrow 0.$$

The module L of first-order syzygies of E, may be identified to the dual module of the first order syzygies of the ideal $I = I_1(\psi)$. Now $(\wedge^t L)^{**} \cong Z_{m-t-1}$, where the Z_i denote the cycles of the ordinary Koszul complex defined by the ideal I.

In this section we shall describe several relationships between I and the ideal $J(E)$ of 1-forms defined by φ in the polynomial ring $B = S(R^n) = R[T_1, \ldots, T_n]$.

Lemma 2.4.1 ([54]) *Assume E satisfies condition $\mathcal{F}_1$. Then:*

(a) *I satisfies $\mathcal{F}_1$.*

(b) *The converse holds if E is a torsion-free module and E^*, the R-dual of E, is a third syzygy module.*

The proof actually gives a way of constructing, out of an ideal I of height at least three, a module E with the stated properties.

Definition 2.4.2 An ideal I is strongly Cohen-Macaulay (SCM for short) if the Koszul homology of I is Cohen-Macaulay—i.e. for a set x of generators of I (any set would do) the homology modules of the Koszul complex $K(x)$, built on x, are Cohen-Macaulay.

Theorem 2.4.3 ([54]) *Let R be a Cohen-Macaulay integral domain and let E be a torsion-free module as above, such that E^* is a third syzygy module. If I be a strongly Cohen-Macaulay ideal of height three satisfying $\mathcal{F}_1$, then J is a Cohen-Macaulay prime ideal.*

The following puts constraints on E in order for $S(E)$ to be a domain.

Theorem 2.4.4 ([54]) *Let $S(E)$ be an almost complete intersection.*

(a) *If $S(E)$ is a domain, then height (I) is odd.*

(b) *Assume R is a Gorenstein integral domain. If the complex $\mathcal{Z}(E)$ is acyclic and $S(E)$ is a Cohen–Macaulay integral domain, then I is a strongly Cohen–Macaulay ideal of height 3. In addition, the Cohen–Macaulay type of $S(E)$ is $1+$ the Cohen–Macaulay type of I.*

Example 2.4.5 Given a regular local ring (R, m) of dimension $n \geq 4$, Vetter [59] constructs an indecomposable vector bundle on the punctured spectrum of R, of rank $n - 2$. Its module of global sections E has a presentation:

$$0 \longrightarrow R \xrightarrow{\psi} R^n \xrightarrow{\varphi} R^{2n-3} \longrightarrow E \longrightarrow 0.$$

We look at $S(E)$; since E satisfies $\mathcal{F}_1$, by Theorem 1.1.1 $\dim S(E) = 2n - 2$. As $S(E)$ is an almost complete intersection, we can apply to it the result above: For n even, $S(E)$ cannot be an integral domain, while for n odd, and larger than 3, the approximation complex $\mathcal{Z}(E)$ will not be exact. For $n = 5$ the matrix φ is:

$$\begin{bmatrix} -x_2 & x_1 & 0 & 0 & 0 \\ -x_3 & 0 & x_1 & 0 & 0 \\ -x_4 & x_3 & -x_2 & x_1 & 0 \\ -x_5 & x_4 & 0 & -x_2 & x_1 \\ 0 & -x_5 & -x_4 & x_3 & x_2 \\ 0 & 0 & -x_5 & 0 & x_3 \\ 0 & 0 & 0 & -x_5 & x_4 \end{bmatrix}.$$

The ideal of definition of $S(E)$ is $J(E) = [T_1, \cdots, T_7] \cdot \varphi$. An application of the *Macaulay* program yielded that $J(E)$ is a Cohen-Macaulay prime ideal. *Macaulay* was again used to show that, for $n = 7$, $S(E)$ is not an integral domain.

2.5 Modules of projective dimension two

The results of the previous section already indicate the difficulties of ascertaining when the symmetric algebra of a module of projective dimension two is a domain—or, Cohen–Macaulay. There are however many interesting modules in this situation, so that this hurdle must be faced.

Before we look at a family of cases with many unresolved questions, let us add another example where fortuitous elements play a role. It is, like Theorem 2.4.4, rather extremal in its hypotheses.

Theorem 2.5.1 ([52, Theorem 3.5]) *Let R be a Cohen–Macaulay integral domain and let E be torsion-free R-module with a resolution*

$$0 \longrightarrow R^{m-2} \xrightarrow{\psi} R^m \xrightarrow{\varphi} R^n \longrightarrow E \longrightarrow 0.$$

If E satisfies $\mathcal{F}_1$, the presentation ideal $J(E)$ is a Cohen–Macaulay prime ideal and is defined as the ideal generated by the $m-1$ sized minors of a matrix gotten by adding to ψ a column of linear forms.

The proof essentially builds the aforementioned forms to be added to ψ.

Let us look, with some detail, at the Jacobian modules of an important class of algebras. Let L be a semisimple Lie algebra over an algebraically closed field k, of characteristic zero. Denote by R ring of polynomial functions on L, $R = k[x_1, \ldots, x_n]$, $n = \dim L$. Let $\mathcal{L} = L \otimes R$. Let $B(,)$ denote the Killing form of L; we extend it to $\mathcal{L}$. To make phenomena of skew symmetry more visible, we may choose a base of L with $B(e_i, e_j) = \delta_{ij}$. Since adx is now skew symmetric one has the following¿

Proposition 2.5.2 *The Jacobian module of L gives rise to the exact sequence*

$$0 \longrightarrow \mathcal{C} \xrightarrow{\psi} \mathcal{L} \xrightarrow{\varphi} \mathcal{L} \longrightarrow E \longrightarrow 0,$$

where $\varphi(a) = [x, a]$, x a generic element of L.

One of the main theorems about these modules will say that $\mathcal{C}$ is a free module and describe its generators.

If $f(x)$ is a polynomial function on L, define its *gradient*, $\nabla f(x)$, by

$$B(\nabla f(x), y) = df_x(y).$$

It follows that, since B is unimodular, $\nabla f(x)$ is an element of $\mathcal{L}$.

Let G be the group of inner automorphisms of L.

Proposition 2.5.3 (N. Wallach) *Let $p(x)$ be an invariant polynomial under G. Then*

$$[\nabla p(x), x] = 0.$$

Proof. If ady is nilpotent, then

$$p(x) = p(e^{t\,ady}x) = p(x + t[x, y] + O(t^2)).$$

Hence $dp_x([y, x]) = 0$ for y nilpotent. But as L is the span of the nilpotent elements, $dp_x([x, L]) = 0$.
Now if $y \in L$,

$$B([\nabla p(x), x], y) = B(\nabla p(x), [x, y]) = dp_x([x, y]) = 0.$$

Thus $[\nabla p(x), x] = 0$. $\square$

This identifies sufficiently many elements to generate $\mathcal{C}$:

Theorem 2.5.4 *Let L be a semisimple Lie algebra of rank ℓ, over an algebraically closed field of characteristic zero. Let $p_1, \ldots, p_\ell$ be homogeneous polynomials generating the subring of invariants of L. The subalgebra C is generated as an R–module by $\nabla p_1, \ldots, \nabla p_\ell$.*

Proof. It has the structure of the proof of [6, Theorem 2.2.1], so we only give the highlights. Let C_0 be the submodule of C spanned by ∇p_i, $1 \leq i \leq \ell$. To prove the equality $C_0 = C$ we use the criterion of [8] (see also [6]). It will suffice to show that C_0 is a free submodule of $\mathcal{L}$, of rank ℓ, and that the ideal generated by the $\ell \times \ell$ minors of the embedding $C_0 \hookrightarrow \mathcal{L}$ has codimension at least two. This is just the Jacobian ideal of the collection of invariant polynomials.

Both conditions follow from a theorem of Kostant [34, Theorem 0.8], who proved that the cone $V(p_1, \ldots, p_\ell)$ is a normal, complete intersection of codimension ℓ. $\square$

What is not clear is the relationship between the projective dimension of the Jacobian module and the structure of the Lie algebra. It is not the case that proj dim $(E) = 2$ characterizes reductive Lie algebras.

As an application we have a very explicit description of the ring $D(L)$ of regular functions on $C(L)$.

Corollary 2.5.5 *Let L be a semisimple Lie algebra of rank ℓ, and invariant polynomials $p_1, \ldots, p_\ell$ as above. Let $T_1, \ldots, T_\ell$ be a fresh set of indeterminates. Consider the element*

$$z = \sum_{i=1}^{i=\ell} p_i T_i \in S = R[T_1, \ldots, T_\ell].$$

$D(L)$ is the R–subalgebra of S generated by the components of the gradient of z.

Remark 2.5.6 The theorem provides the means for computing the invariant polynomials, at least when L is an algebra of low rank. The point is that programs such as *Macaulay* will determine the syzygies of φ, that is, the $\nabla p(x)$, with $p(x)$ recovered through Euler's formula

$$\deg(p) \cdot p(x) = B(\nabla p(x), x).$$

It is straightforward to set up the computation.

To illustrate, in the case of an algebra of type G_2, using its 'most' natural basis (*cf.* [25])—not orthonormal—one obtains for the generic Cartan subalgebra, in addition to x, another vector with fifth degree components $(b_1, \ldots, b_{14})$. They are however practically dense polynomials. Here, for instance is b_1 taken from a *Macaulay* session. The second invariant polynomial p_2 can be found by the method indicated earlier; it is about 30 times longer. (A reduction to an orthonormal basis was carried out but it is not advisable.)

```
-x[2]3x[3]x[5]+1/2x[2]x[3]2x[5]2+x[2]3x[4]x[7]-x[2]2x[3]x[6]x[7]
+1/2x[3]2x[5]x[6]x[7]-1/2x[2]x[4]2x[7]2+1/2x[3]x[4]x[6]x[7]2
-x[2]2x[4]x[5]x[8]+1/2x[3]x[4]x[5]2x[8]-x[2]x[3]x[5]x[6]x[8]
+1/2x[4]2x[5]x[7]x[8]+x[2]x[4]x[6]x[7]x[8]-x[3]x[6]2x[7]x[8]
-x[4]x[5]x[6]x[8]2+6x[2]x[4]x[9]2x[10]-6x[3]x[6]x[9]2x[10]
-3x[4]x[5]x[9]x[10]2+3x[5]x[6]x[10]3+6x[2]x[3]x[9]2x[11]
```

+6x[4]x[8]x[9]2x[11]+3x[3]x[5]x[9]x[10]x[11]
-3x[4]x[7]x[9]x[10]x[11]-6x[2]x[5]x[10]2x[11]+3x[6]x[7]x[10]2x[11]
+3x[3]x[7]x[9]x[11]2-6x[2]x[7]x[10]x[11]2-3x[5]x[8]x[10]x[11]2
-3x[7]x[8]x[11]3-3x[2]x[3]x[5]x[9]x[12]+3x[2]x[4]x[7]x[9]x[12]
-3x[3]x[6]x[7]x[9]x[12]-3x[4]x[5]x[8]x[9]x[12]
+4x[2]2x[5]x[10]x[12]-1/2x[3]x[5]2x[10]x[12]
-1/2x[4]x[5]x[7]x[10]x[12]+4x[5]x[6]x[8]x[10]x[12]
+4x[2]2x[7]x[11]x[12]-1/2x[3]x[5]x[7]x[11]x[12]
-1/2x[4]x[7]2x[11]x[12]+4x[6]x[7]x[8]x[11]x[12]
+4x[2]2x[3]x[9]x[13]-1/2x[3]2x[5]x[9]x[13]
-1/2x[3]x[4]x[7]x[9]x[13]+4x[3]x[6]x[8]x[9]x[13]+x[2]3x[10]x[13]
-2x[2]x[3]x[5]x[10]x[13]-5x[2]x[4]x[7]x[10]x[13]
+3/2x[3]x[6]x[7]x[10]x[13]+3/2x[4]x[5]x[8]x[10]x[13]
+x[2]x[6]x[8]x[10]x[13]+x[2]2x[8]x[11]x[13]
-7/2x[3]x[5]x[8]x[11]x[13]-7/2x[4]x[7]x[8]x[11]x[13]
+x[6]x[8]2x[11]x[13]-9x[2]x[9]x[10]x[12]x[13]
+9/2x[5]x[10]2x[12]x[13]-9x[9]x[8]x[9]x[11]x[12]x[13]
+9/2x[7]x[10]x[11]x[12]x[13]-6x[2]x[7]x[12]2x[13]
+6x[5]x[8]x[12]2x[13]+9/2x[3]x[9]x[10]x[13]2-9/2x[2]x[10]2x[13]2
-9/2x[8]x[10]x[11]x[13]2+3x[3]x[7]x[12]x[13]2-3x[3]x[8]x[13]3
+4x[2]2x[4]x[9]x[14]-1/2x[3]x[4]x[5]x[9]x[14]
-1/2x[4]2x[7]x[9]x[14]+4x[4]x[6]x[8]x[9]x[14]+x[2]2x[6]x[10]x[14]
-7/2x[3]x[5]x[6]x[10]x[14]-7/2x[4]x[6]x[7]x[10]x[14]
+x[6]2x[8]x[10]x[14]-x[2]3x[11]x[14]+5x[2]x[3]x[5]x[11]x[14]
+2x[2]x[4]x[7]x[11]x[14]+3/2x[3]x[6]x[7]x[11]x[14]
+3/2x[4]x[5]x[8]x[11]x[14]-x[2]x[6]x[8]x[11]x[14]
-9x[6]x[9]x[10]x[12]x[14]+9x[2]x[9]x[11]x[12]x[14]
+9/2x[5]x[10]x[11]x[12]x[14]+9/2x[7]x[11]2x[12]x[14]
-6x[2]x[5]x[12]2x[14]-6x[6]x[7]x[12]2x[14]
+9/2x[4]x[9]x[10]x[13]x[14]-9/2x[6]x[10]2x[13]x[14]
+9/2x[3]x[9]x[11]x[13]x[14]-9/2x[8]x[11]2x[13]x[14]
-3x[3]x[5]x[12]x[13]x[14]+3x[4]x[7]x[12]x[13]x[14]
+6x[2]x[3]x[13]2x[14]-3x[4]x[8]x[13]2x[14]+9/2x[4]x[9]x[11]x[14]2
-9/2x[6]x[10]x[11]x[14]2+9/2x[2]x[11]2x[14]2-3x[4]x[5]x[12]x[14]2
+6x[2]x[4]x[13]x[14]2+3x[3]x[6]x[13]x[14]2+3x[4]x[6]x[14]3

These modules have resolutions

$$0 \longrightarrow R^\ell \longrightarrow R^n \overset{\varphi}{\longrightarrow} R^n \longrightarrow E \longrightarrow 0$$

with φ a skew symmetric matrix. It makes them similar to Gorenstein ideals of codimension three. This is further enhanced by the following result of Richardson [45]. (The case of $n \times n$ matrices was proved earlier by Motzkin and Tausski [43] and Gerstenhaber [16].)

Theorem 2.5.7 *The commuting variety of a semisimple Lie algebra is irreducible.*

Several consequences for modules such as those of Theorem 2.5.4 are interesting. Here is a sample:

Proposition 2.5.8 *Suppose $Spec(S(E))$ is irreducible; if E has projective dimension 2 then E is torsion-free.*

Proof. Let

$$0 \longrightarrow R^\ell \longrightarrow R^m \overset{\varphi}{\longrightarrow} R^n \longrightarrow E \longrightarrow 0$$

be a projective resolution of E. If P is an associated prime ideal of E, it must have height at most two (cf. [40, Theorem 19.1]). We claim that $P = (0)$.

Denoting still by R the localization at P, we assume that the resolution above is minimal. Since E satisfies the condition $\mathcal{F}_1$, height $I_1(\varphi) \geq (m - \ell) - 1 + 2$; thus if E is not free, height $P = 2$ and $m - \ell = 1$. This means that there is an exact sequence

$$0 \longrightarrow I \longrightarrow R^r \longrightarrow E \longrightarrow 0,$$

where I is a rank one, non-free module and E is free in codimension one; I may be identified to an ideal of height two. We claim that the symmetric algebra $S(E)$ has at least 2 minimal primes. If $I = (a_1, \ldots, a_s)$, denote by f_i the image of a_i in $R^r = RT_1 \oplus \cdots \oplus RT_r$. The symmetric algebra of E is $R[T_1, \ldots, T_r]/(f_1, \ldots, f_s)$. Since height $(I) = 2$, for any a_i there exists an element $b_i \in I$ such that $\{a_i, b_i\}$ is a regular sequence. If we denote by g_i the image of b_i, we must have $a_i g_i = b_i f_i$ since they are both the image of the element $a_i b_i$. This implies that f_i must be a multiple of a_i. Thus $(f_1, \ldots, f_s) = (If)$, for some 1-form f. $\square$

The meaning is that the defining ideal $J(E)$ of $S(E)$, and its radical, can only begin to differ in degree 2 in the T_i's–variables.

Problem 2. Is the symmetric algebra of the Jacobian module of a semisimple Lie algebra always an integral domain?

Example 2.5.9 We inject a word of caution regarding the primeness of the ideal $J(A)$. Let A be the algebra of *Cayley* numbers. This is the algebra of pairs of quaternions, written $q + re$, with multiplication

$$(q + re)(s + te) = (qs - \bar{t}r) + (tq + r\bar{s})e.$$

Its commuting variety (properly complexified), has the following properties.

The base ring is $R = k[x_1, \ldots, x_8]$; the Jacobian module $E = R \oplus F$ where F has a resolution

$$0 \longrightarrow R \longrightarrow R^7 \overset{\varphi}{\longrightarrow} R^7 \longrightarrow F \longrightarrow 0$$

where φ is a skew symmetric matrix. It satisfies $\mathcal{F}_0$ but not $\mathcal{F}_1$.

$J(A)$ is a Cohen–Macaulay ideal and an almost complete intersection of height 6; by Proposition 2.1.1, it cannot be a prime ideal. It likely has two irreducible components.

We give an example in prime characteristic.

Example 2.5.10 Let W_1 be the Lie algebra that has for basis

$$\{e_1 \mid i \in \mathbb{Z}/(p)\}$$

and multiplication

$$[e_i, e_j] = (j - i)e_{i+j}.$$

For $p = 5$, the Jacobian module of W_1 satisfies the condition $\mathcal{F}_0$ but not $\mathcal{F}_1$ and therefore $C(W_1)$ is not irreducible (the ideal $J(W_1)$ is Cohen–Macaulay). There are additional marked differences between this case and those in characteristic zero. Some can be traced to the non-existence of an invariant non-degenerate quadratic form. In the example, this shows up in the fact that the kernels of the Jacobian matrix and its transpose have different degrees.

2.6   Approximation complexes

Let R be a ring and E a finitely generated R-module. Suppose that

$$\mathcal{F} : 0 \longrightarrow F_n \xrightarrow{f_n} \cdots \longrightarrow F_2 \xrightarrow{f_2} F_1 \xrightarrow{f_1} F_0$$

is a projective resolution of E. In [36] and [62] it is constructed associated complexes of free modules over the symmetric and exterior powers of E.

The complex over the pth symmetric power of E, $S_p(\mathcal{F})$ is defined as follows (*cf.* [62]). For a sequence $a_0, \ldots, a_n$ of non-negative integers, pu

$$S(a_0, \ldots, a_n; \mathcal{F}) = \wedge^{a_0} F_0 \otimes D_{a_1} F_1 \otimes \wedge^{a_2} F_2 \otimes \cdots,$$

with $D_s(-)$ standing for the sth divided power functor. Now put

$$S_t(\mathcal{F})_r = \bigoplus S(a_0, \ldots, a_n; \mathcal{F})$$

for all sequences with $\sum_i a_i = p$ and $\sum_j j a_j = r$. The mappings are derived from the f_i's.

We shall attempt to pinpoint some general difficulties in the use of these complexes. Let

$$0 \longrightarrow R^p \xrightarrow{\psi} R^m \xrightarrow{\varphi} R^n \longrightarrow E \longrightarrow 0$$

be a free resolution of the module E.

For $S(E)$ to be a domain the symmetric powers $S_t(E)$ must be torsion–free modules. Let us use the complex of [62]; for $t = 2$ one has the complex $\mathcal{C}$:

$$0 \to D_2(R^p) \to R^p \otimes R^m \to R^p \otimes R^n \oplus \wedge^2 R^m \to R^m \otimes R^n \to S_2(R^n) \to 0.$$

For $\mathcal{C}$ to be acyclic—*in the presence of* $\mathcal{F}_1$—requires height$(I_p(\psi)) \geq 4$, *cf.* [62]. Thus if in addition E is a reflexive module, it will be satisfied. Consequently we obtain:

Proposition 2.6.1 *Let R be a four-dimensional domain and let E be a reflexive module of projective dimension two. Then $S(E)$ is not an integral domain.*

Two comments are relevant here. If E is a module of projective dimension 2 or higher, the complexes $S_t(\mathcal{F})$ cannot be exact for t large. Nevertheless, the complexes, for low values of p, can be still used to feed information into another family of complexes over the symmetric powers of E.

For the theory of the approximation complexes, we refer the reader to [20] and [21]. We only recall the definition of the modified Koszul complex of a module [21, page 668]. First, the Koszul complex associated to the presentation of the module E:

$$0 \longrightarrow Z_1(E) = L \xrightarrow{\varphi} R^n = F \longrightarrow E \longrightarrow 0,$$

is defined as $\mathcal{K}(E) = \wedge(L) \otimes S(F)$ with differential

$$\partial((a_1 \wedge \cdots \wedge a_r) \otimes w) = \sum (-1)^j (a_1 \wedge \cdots \wedge \hat{a}_j \wedge \cdots \wedge a_r) \otimes \varphi(a_j) \cdot w.$$

Assume E is a torsion–free module. If we replace each $\wedge^r L$ by its bi-dual $Z_r = Z_r(E) = (\wedge^r L)^{**}$, we obtain the $\mathcal{Z}(E)$ complex of E $(B = S(F))$:

$$0 \to Z_\ell \otimes B[-\ell] \to \cdots \to Z_1 \otimes B[-1] \to B \longrightarrow S(E) \to 0.$$

Here $\ell = \text{rank}(E)$, and the complex is just a generalization, with non-free components, of $S(\mathcal{F})$, for modules of projective dimension one. It is easy to see that they agree in this case.

Observe that the complex $\mathcal{Z}(E)$ is, in general, reasonably short. To be useful, depth information about the $Z_i(E)$ should be available.

The following result summarizes several aspects of these complexes.

Theorem 2.6.2 ([21]) *Let R be a Noetherian local ring with infinite residue field.*

1. *The following are equivalent:*

    (a) $\mathcal{Z}(E)$ *is acyclic.*

    (b) S_+ *is generated by a d–sequence of linear forms of $S(E)$.*

2. *If $\mathcal{Z}(E)$ is acyclic, the Betti numbers of $S(E)$ as a module over $B = S(R^n)$ are given by*
$$\beta_i^B(S(E)) = \sum_j \beta_j^R(Z_{i-j}(E)).$$

3. *If R is Cohen–Macaulay and E has rank e, the following conditions are equivalent:*

    (a) $\mathcal{Z}(E)$ *is acyclic and $S(E)$ is Cohen–Macaulay.*

    (b) E *satisfies $\mathcal{F}_0$ and*
$$\text{depth } Z_i(E) \geq d - n + i + e, \ i \geq 0.$$

4. *Moreover, if R is Cohen–Macaulay with canonical module ω_R then*

    (a)
$$\omega_S/S_+\omega_S = \oplus_{i=0}^\ell \text{Ext}^{\ell-i}(Z_i(E), \ \omega_R).$$

    (b) $S(E)$ *is Gorenstein if and only if $\text{Hom}_R(Z_{n-e}(E),\omega_R) = R$ and*
$$\text{depth } Z_i(E) \geq d - n + i + e + 1, \ i \leq n - e - 1.$$

A significant point of the construction above is its length, equal to the torsion–free rank of L. If, for instance, E satisfies $\mathcal{F}_k$ then $\text{rank}(L) \leq \dim R - k$. In general, if E satisfies $\mathcal{F}_0$ the ideal of definition of $S(E)$ is height unmixed.

Algebras of low codimension, that is when $\text{rank}(L) \leq 4$ are easier to analyze. In the simplest case, when E is torsion-free and $\text{rank}(L) = 2$, we have the approximation complex:

$$0 \to (\wedge^2 L)^{**} \otimes S_{t-2}(R^n) \to L \otimes S_{t-1}(R^n) \to S_t(R^n) \to S_t(E) \to 0.$$

If E has projective dimension finite, not necessarily two—or, if R is a factorial domain—$(\wedge^2 L)^{**} \cong R$. It follows that if E satisfies $\mathcal{F}_1$ then $S(E)$ is a domain. In fact, if R is a Cohen-Macaulay ring, then $S(E)$ is a codimension two Cohen-Macaulay domain if and only if L has projective dimension at most one. Notice how this argument even strengthens Theorem 2.5.1.

This short complex typifies some of the differences between the approximation complexes and projective resolutions of symmetric powers.

If $\mathrm{rank}(L) = 3$, under similar conditions, the approximation complex is still acyclic. To get depth information we need to find out the depth of $(\wedge^2 L^{**})$. By duality this module is isomorphic to L^*. Let us examine a simple case.

Proposition 2.6.3 *Assume that the module E is torsion-free and has a presentation*

$$0 \longrightarrow C \overset{\psi}{\longrightarrow} R^n \overset{\varphi}{\longrightarrow} R^n \longrightarrow E \longrightarrow 0,$$

where φ is either symmetric or skew symmetric. Then $L \simeq L^*$.

Proof. Dualizing the presentation we get that the image of φ^*—which we can identify to L—embeds into the kernel of ψ^*—which is L^*. Since this embedding of reflexive modules is an isomorphism in codimension one, they must be isomorphic. $\square$

If additionally the module E has projective dimension two and satisfies $\mathcal{F}_1$, then an application of Theorem 2.6.2 will say that $S(E)$ is a Cohen-Macaulay domain.

Problem 3. Find necessary and sufficient conditions for the integrality of $S(E)$ in codimensions three and four.

3 Jacobian criteria

This section is concerned with the regular prime ideals of a symmetric algebra $S = S(E)$.

It begins with the identification of some of regular primes of a symmetric algebra $S(E)$, and as an application has the proof of the *Zariski–Lipman conjecture* for symmetric algebras over regular rings. This relative version is much simpler than the other absolute cases of the conjecture that have been established.

A normality test that is easy to apply is given next. It still requires that the condition S_2 of Serre be detected by other means. Normal algebras over factorial domains have freely generated divisor class groups, of rank given by a formula read off the presentation of the module.

3.1 Regular primes

There is a set of such primes that is easy to deal with:

Proposition 3.1.1 *Let $Q \supseteq S_+$ be a regular prime of S; put $P = Q \bigcap R$. Then R_P is a regular local ring and E_P is a free R_P-module.*

Proof. After localizing at P, Q becomes the irrelevant maximal ideal of the graded algebra S_P. Since its embedding dimension, $v(P) + v(E_P)$, must equal the Krull dimension of S_P, it is clear that P and E are as asserted. $\square$

The following syzygy theorem is immediate:

Corollary 3.1.2 *Let R be a Noetherian ring of finite Krull dimension and let E be a finitely generated R-module. Then $S(E)$ has finite global dimension if and only if R is regular and E is a projective R-module.*

The next elementary result describes the module $\Omega_{S/R}$ of relative differentials.

Proposition 3.1.3 $\Omega_{S/R} \simeq E \otimes_R S$.

Proof. Write $S = B/J$, where B is a polynomial ring $R[T_1, \ldots, T_n]$. The exact sequence of modules of differentials,

$$J/J^2 \longrightarrow \Omega_{B/R} \otimes_B S \longrightarrow \Omega_{S/R} \longrightarrow 0$$

is precisely the presentation of E over R tensored by S. $\square$

The following relative version of the *Zariski–Lipman conjecture* is inspired by [38] and [24].

Theorem 3.1.4 *Let R be an affine algebra over a field k of characteristic zero, and let E be a finitely generated R-module. If the module of k-derivations $\mathcal{D} = Der_k(S(E), S(E))$ is a projective $S(E)$-module, then $S(E)$ is a smooth R-algebra.*

Proof. It consists of several steps. To begin we localize R at a maximal ideal; we shall prove that E is a free R-module.

Step 1. $\mathcal{D}$ is a graded S-module ($S = S(E)$). Indeed, if we present

$$S = B/J = R[T_1, \ldots, T_n]/(f_j = \sum a_{ij}T_i, \ j = 1, \ldots, m)$$

the module $\Omega_{S/k}$ is obtained as follows. First, grade the B-module $\Omega_{B/k}$ so that the differentials $d(r)$, $r \in R$, have degree zero, and $d(T_i)$ has degree one. From the fundamental sequence for modules of differentials, $\Omega_{S/k}$ is the quotient of $\Omega_{B/k}$ by the relations

$$\sum d(a_{ij})T_i \ + \ \sum a_{ij}d(T_i) = 0, \ j = 1, \ldots, m,$$

and the equations of $J(E)$. Since these are all homogeneous, $\Omega_{S/k}$ is a graded S-module and therefore its dual $\mathcal{D}$ will also be graded.

Step 2. The inclusion $R \hookrightarrow S$ gives rise to the exact sequence

$$\Omega_{R/k} \otimes S \longrightarrow \Omega_{S/k} \longrightarrow \Omega_{S/R} \longrightarrow 0,$$

which by dualizing and the formula for relative differentials yields

$$0 \longrightarrow Hom_S(E \otimes_R S, S) \longrightarrow \mathcal{D} \longrightarrow Hom_S(\Omega_{R/k} \otimes_R S, S).$$

By adjointness, the module on the left can be written

$$Hom_R(E, S) = Hom_R(E, R) \oplus Hom_R(E, E) \oplus \cdots$$

with the appropriate grading.

From [38] we know that S is a normal domain. Therefore $\dim S = \dim R + \text{rank}(E) = d+e$. Pick, by the previous step, a homogeneous basis $\{D_s, \ s = 1, \ldots, d+e\}$ of $\mathcal{D}$. Assume that D_s, $s \leq r$, are the derivations of degree -1. We claim that when restricted to E, they generate $Hom_R(E, R)$. Suppose otherwise, and let $\varphi \in Hom_R(E, R)$ but not in the span of the D_s, $s \leq r$:

$$\varphi = \sum_{s \leq r} a_s D_s + \sum_{s > r} a_s D_s.$$

But it is clear that if the degree of D_s is not -1 then $a_s = 0$ since S is positively graded.

By dimension counting, $Hom_R(E, S)$ is a module of rank e over S, containing a free summand of rank equal to the R-rank of $Hom_R(E, R)$. Thus $Hom_R(E, S)$ is a free S-module, of rank e.

Step 3. Since $Hom_R(E, S)$ is free on elements of degree -1, we get that $Hom_R(E, R) \simeq R^e$ and $Hom_R(E, E) \simeq E^e$. We claim that E is R-free. Note that R is a (normal) domain and E is torsion-free ($E \subset S$), and we may assume that E is free on the punctured spectrum of R, with $\dim R > 1$.

We recall the canonical mapping

$$E \otimes_R Hom_R(E, R) \xrightarrow{\psi} Hom_R(E, E)$$

defined by

$$\psi(e \otimes f)(x) = f(x) \cdot e, \ \text{for} \ e, x \in E, \ f \in Hom_R(E, R).$$

According to [2, Proposition A.1], E is R–projective if and only if ψ is surjective.

With the identifications above, ψ is an endomorphism of E^e:

$$E^e \xrightarrow{\psi} E^e.$$

Because E is torsion-free and $Hom_R(E, R)$ is R–free, this map is injective. Denote the cokernel by H. It suffices to prove that $H = 0$; this will follow from the next lemma.

Lemma 3.1.5 *Let R be a Noetherian ring and let G be a finitely generated R-module with $\dim G = \dim R$. Then any injective endomorphism α of G, which is an isomorphism in codimension at most one, is an isomorphism.*

Proof. Consider the exact sequence

$$0 \longrightarrow G \xrightarrow{\alpha} G \longrightarrow H \longrightarrow 0.$$

By induction we may assume that R is a ring of dimension at least 2, and H is a module of finite length. We replace R by the polynomial ring $R[t]$ modulo the characteristic polynomial of α, while preserving all the other hypotheses—that is, we may assume that α is multiplication by an element of the ring. But then if α is not a unit, $\dim H \geq \dim G - 1$ according to [40, 12.F]. □

3.2   Normality

A normality criterion for $S(E)$ is given in [54]. We assume that $S = S(E)$ is an integral domain. Let

$$0 \longrightarrow L \longrightarrow R^n \longrightarrow E \longrightarrow 0,$$

be a minimal presentation of E. Because of $\mathcal{F}_1$ we have $\ell = \text{rank}\,(L) \leq \dim R - 1$.

To apply Serre's normality criterion we assume that S satisfies condition S_2 and examine the localizations S_Q where Q is a prime of height 1. If $Q \cap R = P = (0)$, there is nothing to do since S_P is a polynomial ring over the field of fractions of R. If $P \neq (0)$, localize at P and again let R stand for R_P. We thus have that $Q = PS$. Since $\dim S/PS = v(E) = \dim S - 1$, we get $\ell = \text{rank}(L) = \text{height}(P) - 1$.

Denote by B the polynomial ring $S(R^n) = R[T_1, \ldots, T_n]$ and let J be the ideal of linear forms generated by the images of the elements of L in B. Let $M = PB$. We now convert to B the condition that S_Q be a discrete valuation domain. This is the case if and only if $(M/J)_M$ is a principal ideal, that is, if and only if the image of J in $(M/M^2)_M$ has rank $edim\,(R) - 1$.

Now we rephrase this last condition into a more visible criterion. Let $f_j = (a_{ij})$, $j = 1, \ldots, m$, be a minimal generating set of L. If we choose a minimal set of generators of the maximal ideal $P = \{x_1, \ldots, x_r\}$, we may write each a_{ij} as a linear combination of the x_k,

$$a_{ij} = \sum_{k=1}^{k=r} a_{ij}^{(k)} x_k.$$

The matrix $\varphi = (a_{ij})$ can be written as the product of two matrices, $A = [A_1, \cdots, A_m]$ and $U(X)$, where A_j is the matrix block $(a_{ij}^{(k)})$ and $U(X)$ is the $m \cdot r \times r$ matrix

$$\begin{bmatrix} X & 0 & \cdots & 0 \\ 0 & X & \cdots & 0 \\ \vdots & \vdots & \ddots & \vdots \\ 0 & 0 & \cdots & X \end{bmatrix}$$

made up of blocks of $[x_1, \ldots, x_r]$.

Proposition 3.2.1 *Let R be an integral domain, and let E be a finitely generated R-module whose symmetric algebra algebra $S(E)$ is an integral domain with the Serre's condition S_2. $S(E)$ is normal if and only if for each prime ideal P of R such that $v(E_P) - \text{rank}\,(E) = \text{height}(P) - 1$, the rank of the matrix*

$$J(\varphi) = [\,^t A_1^*, \ldots, \,^t A_m^*] \cdot U(T)$$

is equal to $edim(R_p) - 1$. Here $^t A_j^*$ denotes the transpose of A_j with its entries taken modulo P, and $U(T)$ is the analog in the T_i's of the matrix $U(X)$.

Proof. The ideal J is generated by the forms

$$\sum_i a_{ij} T_i = \sum (\sum a_{ij}^{(k)} T_i) x_k, \quad j = 1, \ldots, m,$$

so that $J(\varphi)$ defines the image of J into the vector space $(M/M^2)_M$. $\square$

Note that when R is factorial and $S(E)$ is an integral domain satisfying the S_2 condition of Serre, Theorem 3.3.4 describes the finite set of primes that must be tested.

Let us apply this criterion to Example 2.4.5. Because E is free on the punctured spectrum, we only have to verify the criterion at the maximal ideal. The Jacobian matrix of $J(E)$ with respect to the x-variables is:

$$J(\varphi) = \begin{bmatrix} 0 & T_1 & T_2 & T_3 & T_4 \\ -T_1 & 0 & -T_3 & -T_4 & T_5 \\ -T_2 & T_3 & 0 & T_5 & T_6 \\ -T_3 & T_4 & -T_5 & 0 & T_7 \\ -T_4 & -T_5 & -T_6 & -T_7 & 0 \end{bmatrix}.$$

Since rank $J(\varphi) = 4$, $S(E)$ is integrally closed.

For another example, consider a complete intersection (*cf.* [48]):

Example 3.2.2 Let R be a 3-dimensional regular local ring, and let E be the module defined by the matrix

$$\begin{bmatrix} a & 0 \\ b & a \\ 0 & b \\ c & 0 \\ 0 & c \end{bmatrix}$$

where $\{a, b, c\}$ is a regular sequence in R. E is free on the punctured spectrum of R and $S(E)$ is a domain and a complete intersection. By the criterion it follows that $S(E)$ is normal if and only if $\{a, b, c\}$ contains at least two independent minimal generators of the maximal ideal of R.

3.3   Divisor class group

Let R be an integral and suppose E is a finitely generated module such that $S = S(E)_{red}$ is an integrally closed domain. Let

$$R^m \xrightarrow{\varphi} R^n \longrightarrow E \longrightarrow 0$$

be a presentation of E. We intend to show that the cardinality of the divisor class group of S is related to various sets of associated prime ideals. A broader version of it appears in [22] (see also [52]).

Assume that $T(P)$ is a prime ideal of height 1. We have

$$\dim S/T(P) = \dim S - 1 = \dim R + \text{rank}(E) - 1 = d + e - 1.$$

On the other hand, assume that E is minimally generated by $n - t$ elements at P, $t \geq 0$. in view of Theorem 1.1.1,

$$\dim S/T(P) = \dim R/P + v(E_P),$$

from which we get

$$\text{height}(P) = n - t - e + 1.$$

The condition $v(E_P) = n - t$ implies that $I_t(\varphi) \not\subset P$, but $I_{t+1}(\varphi) \subset P$. Because of the $\mathcal{F}_1$–condition on E, this implies that

$$\text{height}(P) \geq (n - e) - (t + 1) + 2,$$

and therefore we must have that P is a minimal prime of $I_{t+1}(\varphi)$, of height $n - e - t + 1$.

Of course this argument breaks down when $t = n - e$, but then the localization E_P is a free module.

Definition 3.3.1 Let E be an R-module satisfying $\mathcal{F}_1$ with a presentation as above. For each integer $1 \leq t \leq n - e$, denote by h_t the number of minimal primes of $I_t(\varphi)$ of height $n - e - t + 2$. The sum

$$\sum_{t=1}^{n-e} h_t$$

is the f–number of E.

Remark 3.3.2 The f–designation somewhat unappropriately stands for Fitting. Because A. Simis has, to a greater extent than anyone else, searched and computed these numbers, the terminology *Simis–number* would be more fitting.

Unfortunately it is difficult to find directly the minimal primes of such huge–meaning determinantal–ideals. We must often look for indirect means to count these primes.

Definition 3.3.3 Let R be an integral domain and le E be a finitely generated R-module. E is said to be of *analytic type* if $Spec(S(E))$ is irreducible.

The advantage of this definition (*cf.* [6]) is that the condition $\mathcal{F}_1$ still holds for E and in many cases it is much easier to prove irreducibility than integrality. Note that both imply $\mathcal{F}_1$ for E, and the former is (in the presence of this condition) characterized by equi–dimensionality (*cf.* [6]). It would be of interest to express it in ideal-theoretic terms.

Theorem 3.3.4 *Let R be an universally catenarian factorial domain and let E be a module such that $S = S(E)_{red}$ is a normal domain. Then the divisor class group of S is a free abelian group of rank equal to the f–number of E.*

Sketch of the Proof. (See [22] and [60].) If P is a height one prime of S, and $P \cap R = \mathbf{p} \neq (0)$, then $P = T(\mathbf{p})$. If $\mathbf{p} = (0)$, since the localization $S_{\mathbf{p}}$ is a polynomial ring, we can shift the support of P to an isomorphic ideal Q that has a nonzero contraction with R. If $\text{height}(\mathbf{p}) = 1$, $P = T(\mathbf{p})$, with $\mathbf{p} = xR$; this time consider the ideal xP^{-1} to obtain an ideal in the same class as P^{-1} whose contraction to R has height at least two.

This leaves as generators for the divisor class group of S, $Cl(S)$, the classes of $T(\mathbf{p})$, where $\mathbf{p}$ is a prime for which $E_{\mathbf{p}}$ is not a free module. It must then be one of the ideals counted in $f(E)$.

Suppose there is a relation amongst the classes of some of the $T(\mathbf{p}_i)$:

$$\sum_i r_i[T(\mathbf{p}_i)] = 0.$$

In the subset of those primes with $r_i \neq 0$, pick $\mathbf{p} = \mathbf{p}_j$ minimal. Localizing at $\mathbf{p}$, the image of $T(\mathbf{p})$ is $\mathbf{p}S\mathbf{p}$. The divisorial closure of the rth power of such ideal is $\mathbf{p}^r R\mathbf{p} \oplus \cdots$, which cannot be principal as height$(\mathbf{p}) \geq 2$. $\square$

4   Factoriality

The question that drives this section is: When is the symmetric algebra of the R-module E, $S(E)$, factorial? One condition that has been identified is simply $\mathcal{F}_2$ (see below), but it is hardly enough except for modules of projective dimension one. As a matter of fact, all known examples are complete intersections.

If symmetric algebras that are factorial seem rare, there is a straightforward process that produces the *factorial closure* of any symmetric algebra $S(E)$. The setting is a sequence of modifications of the algebra $S(E)$, each more drastic than the preceding. Define:

(i) $D(E) = S(E)/\text{mod } R\text{-torsion}$ ($D(E)$ is a domain); (ii) $C(E) = $ integral closure of $D(E)$; (iii) $B(E) = $ graded bi-dual of $S(E)$, that is if $S(E) = \oplus S_t(E)$, then $B(E) = \oplus S_t(E)^{**}$, where (**) denotes the bi-dual of an R-module.

These algebras are connected by a sequence of homomorphisms:

$$S(E) \longrightarrow D(E) \longrightarrow C(E) \longrightarrow B(E).$$

The algebra $D(E)$ is easy to obtain from $S(E)$. The other two algebras, $C(E)$ and $B(E)$, are a different matter. The significance of the algebra $B(E)$ is that it is a factorial domain (*cf.* [10]; see also [21], [48]), although it may fail to be Noetherian ([47]).

4.1   The factorial conjecture

In this section we assume that R is a factorial domain. It looks at ways symmetric algebras turn out to be factorial.

The following theorem of Samuel ([48]) exploits the relationship between the factoriality of a graded ring $A = \oplus_{n \geq 0} A_n$, and the A_0-module structure of the components A_n.

Theorem 4.1.1 *Let $A = \oplus_{n \geq 0} A_n$ be a Noetherian, integral domain. The following conditions are equivalent:*

(a) *A is factorial.*

(b) *A_0 is factorial, each A_n is a reflexive A_0-module and $A \otimes_{A_0} K$ is factorial ($K = $ field of quotients of A_0).*

It justifies the assertion above about the algebra $B(E)$. Let us use it to derive a first necessary condition for a symmetric algebra $S(E)$ to be factorial.

Proposition 4.1.2 *If $S(E)$ is factorial then E satisfies $\mathcal{F}_2$.*

Proof. Let $I_t(\varphi)$ be a Fitting ideal associated to E. We already know that E satisfies $\mathcal{F}_1$, so that each of these ideals has height at least 2. We may then find a prime element $x \in R$ contained in their intersection (possibly after a polynomial change of ring; see [9]). Since each $S_n(E)$ is a reflexive R–module, its reduction modulo x is a torsion-free $R/(x)$–module. Therefore $S_{R/(x)}(E/xE)$ is an integral domain, and from its $\mathcal{F}_1$–condition we get the assertion. □

Theorem 4.1.3 *Let R be a factorial domain, and let E be a module of projective dimension 1. Then $S(E)$ is factorial if and only if E satisfies $\mathcal{F}_2$.*

The result, and its proof, is similar to Theorem 2.3.1.

The nature of a module whose symmetric algebra is factorial remains elusive. Although there are many examples, they all seem to fit a mold. It has led us to formulate:

Conjecture 4.1.4 (Factorial conjecture) *Let R be a regular local ring. If $S(E)$ is factorial domain then proj dim $E \leq 1$, that is, $S(E)$ must be a complete intersection.*

To lend evidence, we give other instances where it holds, and connect it to other conjectures. To inject a word of caution, there are modules whose symmetric algebra is factorial but fail to be complete intersection; the base rings are however not regular and the modules have infinite projective dimension (*cf.* [7]).

We begin with a case where this condition gets somewhat strenghtened.

Theorem 4.1.5 ([54, Theorem 3.1]) *Let R be a regular local ring containing a field, and suppose E is a finitely generated R–module such that the enveloping algebra of $S(E)$, i.e., $S(E)_e = S(E) \otimes_R S(E)$, is an integral domain. Then proj dim $E \leq 1$.*

Proof. $S(E)_e$ is just the symmetric algebra of the 'double module' $E \oplus E$. We may assume that for each non-maximal prime ideal $\mathbf{p}$, $E_{\mathbf{p}}$ has projective dimension at most 1 over $R_{\mathbf{p}}$. Let

$$0 \longrightarrow L \longrightarrow R^n \longrightarrow E \longrightarrow 0$$

be a minimal presentation of E. Since $\mathcal{F}_1$ applies to the module $E \oplus E$, we have that ($\ell = \text{rank}(L)$, $d = \dim R$): $2\ell \leq d - 1$.

Let t be the depth of L; because L is free on the punctured spectrum of R, L is a t–syzygy module. We claim that $t \geq \ell + 2$; it will follow from [12] that L must be free. Since $S(E)_e$ is a domain, $E \otimes_R E$ is torsion-free, so that by [1], [37],

$$\text{proj dim } (E \otimes_R E) = 2 \text{ proj dim } E \leq d - 1.$$

But *proj dim E = proj dim $L + 1 = d - t + 1$* and thus $t \geq \ell + 2$. □

Proposition 4.1.6 ([21, Proposition 7.2]) *Let R be a Cohen–Macaulay ring and let E be a finitely generated R–module. If E satisfies $\mathcal{F}_2$ then proj dim $E \neq 2$.*

Proof. Suppose otherwise; pick R local with lowest possible dimension, in particular we may assume *proj dim* $_{R_P}$ $E_P \leq 1$ for each $P \neq M =$ maximal ideal of R. Let

$$0 \longrightarrow R^r \xrightarrow{\psi} R^m \xrightarrow{\varphi} R^n \longrightarrow E \longrightarrow 0$$

be a minimal resolution of E. On account of $\mathcal{F}_2$, we have

$$n = v(E) \leq \dim R + \text{rank } (E) - 2,$$

that is

$$n - r = \ell = \text{rank } (\varphi) = n - \text{rank } (E) \leq \dim R - 2.$$

Since $r \neq 0$, the ideal $I_r(\psi)$ is M–primary. From [11], however, we have

$$\dim R = I_r(\psi) \leq m - r + 1 = \ell - 1,$$

which is a contradiction. $\square$

A few other cases of projective dimension three were resolved in [21], but nothing much beyond is known in the dimension scale.

4.2   Homological rigidity

There seems to be a connection between the conjecture above and another conjecture on the homological rigidity of the module of differentials. Let k be a field of characteristic zero, and let A be a finitely generated k–algebra whose module of differentials, $\Omega_{A/k}$, has finite projective dimension over A. It is conjectured in [56] that A must necessarily be a complete intersection. We follow [58] closely.

To explore this, assume R is a polynomial ring over k, E is a module of projective dimension r, and that $\Omega_{S(E)/k}$ has finite projective dimension over $S(E)$. The validity of the conjecture would imply that $r \leq 1$.

From the exact sequences of modules of differentials

$$0 \longrightarrow \Omega_{R/k} \otimes S(E) \longrightarrow \Omega_{S(E)/k} \longrightarrow \Omega_{S(E)/R} \longrightarrow 0,$$

and the isomorphism $\Omega_{S(E)/R} \simeq E \otimes_R S(E)$ of Proposition 3.1.3, we get that $\Omega_{S(E)/k}$ has finite projective dimension over $S(E)$ if and only if $E \otimes_R S(E)$ does so.

One way to attempt to find a finite projective resolution for $\Omega_{S/R}$ is the following. Let

$$0 \longrightarrow F_r \longrightarrow \cdots \longrightarrow F_1 \longrightarrow F_0 \longrightarrow E \longrightarrow 0$$

be a resolution of E. Tensoring with $S(E)$, we obtain a complex of free $S(E)$–modules over $E \otimes_R S(E)$. It will be a consequence of Proposition 4.2.2 that if $r \geq 2$, this complex is never acyclic.

We begin with an useful feature of symmetric algebras.

Proposition 4.2.1 *Let R be a regular affine domain over a field k of characteristic zero, and let E be a finitely generated R–module. If $\Omega_{S(E)/k}$ has finite projective dimension over $S(E)$, then $S(E)$ is an integral domain.*

Proof. From the earlier remark, the hypothesis is equivalent to proj dim$_S E \otimes S < \infty$.

Since S has no nontrivial idempotent, any module of finite projective dimension over it has a well-defined rank. In this case the rank of $E \otimes S$ must be equal to the R-rank of E. Moreover, by the Auslander-Buchsbaum formula ([40, p. 114]) and the ordinary Jacobian criterion for simple points, S is reduced.

Let Q be a minimal prime of S; put $P = Q \cap R$. Localizing at P, and changing the notation, we may assume that P is the maximal ideal of the local domain R. We claim that $P = 0$; for this it suffices to prove that E is a free R-module.

Let

$$R^m \xrightarrow{\varphi} R^n \longrightarrow E \longrightarrow 0, \quad L = \varphi(R^m),$$

be a minimal presentation of E. Tensoring over with S_Q we obtain a minimal presentation of $E \otimes S_Q$ since the entries of $\varphi \otimes S_Q$ lie in the maximal ideal of S_Q. Therefore rank $(E \otimes S) = n$, which implies $L = 0$ and E is free as asserted. $\square$

The next result works as a mechanism to shrink the projective dimension of E.

Proposition 4.2.2 *Let R be a regular local ring and let E be a finitely generated R-module with proj dim $E = r$. If*

$$Tor_1^R(E, S(E)) = 0,$$

then $r \leq 1$.

Proof. If $r > 1$, we may assume that on the punctured spectrum of R the projective dimension of E is at most 1. The hypothesis means

$$Tor_1^R(E, S_i(E)) = 0, \ i \geq 1,$$

for each of the symmetric powers of E.

We bring in the rigidity of Tor, cf. [1], [37], and the complexes of [62]. We obtain a contradiction by progressively increasing the dimension of R. First, since $Tor_1^R(E, E) = 0$, by [1], [37], the projective dimension of $E \otimes E$ is $2r$, so that dim $R \geq 2r$.

We assume that the complex $\mathcal{F}$ is a minimal resolution of E, and recall the complexes $S_t(\mathcal{F})$ of [62] derived from it, lying over the symmetric powers of E. Their length is given by the formula

$$\begin{cases} nt & \text{for } r \text{ even} \\ inf\{(r-1)t + \text{rank}(F_r), \ nt\} & \text{for } r \text{ odd} \end{cases}$$

By Proposition 4.2.1, E satisfies the condition $\mathcal{F}_1$ of [21]. Since E has projective dimension at most 1 on the punctured spectrum of R, it follows (see [3], [26], [51, Theorem 3.4]) that these complexes have homology concentrated on the maximal ideal of R.

We now repeatedly apply the *lemme d'acyclicité*. For $t = 2$, the length of $S_2(\mathcal{F})$ is at most dim R, so that the complex is acyclic, and therefore a minimal resolution of $S_2(E)$.

Again, from $Tor_1^R(E, S_2(E)) = 0$, we increase the dimension of R enough to guarantee the acyclicity of $S_3(\mathcal{F})$, and so on. $\square$

As an application

Corollary 4.2.3 *Let R be a polynomial ring as above, and assume that proj dim $E = 2$. Then proj dim$_{S(E)}(\Omega_{S(E)/k}) = \infty$.*

Proof. We replace R by a localization ring and assume that E has projective dimension at most 1 on the punctured spectrum of R. Let

$$0 \longrightarrow F_2 \xrightarrow{\psi} F_1 \xrightarrow{\varphi} F_0 \longrightarrow E \longrightarrow 0$$

be a minimal free resolution of E.

If *proj dim* $\Omega_{S/k}$ is finite, S is an integral domain by Proposition 4.2.1 and therefore $Tor_2^R(E, S) = 0$. We then have

$$T = Tor_1^R(E, S) = ker(\varphi \otimes S)/image(\psi \otimes S).$$

The claim is that this module vanishes.

T is a torsion–module of finite projective dimension over S. Furthermore it is graded and annihilated by some power of the maximal ideal of R.

Step 1. To a graded presentation of T:

$$S^p \xrightarrow{\alpha} S^q \longrightarrow T \longrightarrow 0,$$

there is an associated divisorial ideal

$$\mathbf{d}(T) = (I_q(\alpha)^{-1})^{-1}$$

where $I_q(\alpha)$ denotes the ideal generated by the q-sized minors of a matrix representation of α. Because T has finite projective dimension, $\mathbf{d}(T)$ is an invertible ideal of S [39]. As T is graded and R is a local ring, $\mathbf{d}(T)$ must be generated by a homogeneous element $f \in S$. But Sf contains the annihilator of T, in particular a power of the maximal ideal of R. Thus $f \in R$.

Step 2. If f is not a unit of R, dim $R = 1$, which is a contradiction. Otherwise the annihilator of T has grade at least 2, according to [39]. Since both $image(\psi \otimes S)$ and $ker(\varphi \otimes S)$ are second–syzygy modules this is impossible, by standard depth considerations, unless $T = 0$. □

4.3 Finiteness of ideal transforms

To begin our discussion of the factorial closure of a symmetric algebra $S(E)$, we first recall a basic notion of commutative algebra.

Let I be an ideal of the Noetherian integral domain R, of field of fractions K. The *ideal transform* of I, $T_R(I)$, is the subring of all elements of K that can be transported into R by a high enough power of I:

$$B = T_R(I) = \bigcup_i I^{-i}.$$

In other words, B is the ring of global sections of the structure sheaf of R on the open set defined by I. The fundamental reference for this notion is [44]. We point out the following observation. If $I = I_1 \cap I_2$, and I_2 has grade at least two, then the ideal transforms of I and I_1 are the same. Furthermore, if R has the condition S_2 of Serre, we may even replace I by a subideal generated by two elements. As a matter of fact, in all the cases treated here the ideal I can always be taken to be generated by two elements, even when the condition S_2 is not present. This will represent an important computational simplification.

Let R be Cohen-Macaulay factorial domain and let E be a module with a presentation:

$$R^m \xrightarrow{\varphi} R^n \longrightarrow E \longrightarrow 0.$$

Denote by $D(E)$ the quotient of $S(E)$ modulo the ideal of torsion elements. To obtain $B(E)$ one might as well apply the bi-dualizing procedure on $D(E)$. In particular we may assume that E is a torsion-free module. (More about this later.) According to [48], $B(E)$ can be described in the following manner. First, embed $D(E)$ into a polynomial ring $K[U_1, \ldots, U_e]$, K the field of fractions of R and e the rank of E. Then:

$$B(E) = \bigcap_{p \subset R} D(E)_p, \ \text{height}(p) = 1. \tag{0.1}$$

Proposition 4.3.1 *Let J be the Fitting ideal $I_{n-e}(\varphi)$ and put $M = JD(E)$. Then $B(E)$ is the M-ideal transform of $D(E)$.*

Proof. Because E is assumed torsion free, J is an ideal of height at least two. If T denotes the ideal transform of $D(E)$ with respect to J, it is clear from the equation above that $T \subset B(E)$. Conversely, if $b \in B(E) \setminus D(E)$, denote by L the conductor ideal $D(E) :_R b$. Let Q be a minimal prime of L; if J is not contained in Q, the localization E_Q is a free R_Q-module and therefore $S(E_Q) = B(E_Q)$, so that $L_Q = R_Q$—which would be a contradiction. This shows that the radical of L contains J. $\square$

Corollary 4.3.2 *There exists an ideal I generated by a regular sequence $\{f, g\}$ such that $B(E) = T_{D(E)}(I)$.*

Proof. Let $\{f, g\}$ be any regular sequence contained in the ideal J (recall: height$(J) \geq 2$). It is clear that $T_{D(E)}(J) \subseteq T_{D(E)}(I)$. On the other hand, by Equation 0.1 any such transform must be contained in $B(E)$. $\square$

The equality of the algebras $B(E)$ and $D(E)$ has a direct formulation.

Theorem 4.3.3 ([21, Theorem 2.1]) *Let R be a normal, Cohen–Macaulay, universally Japanese domain.*

(a) *If E satisfies $\mathcal{F}_2$ then $B(E) = C(E)$.*

(b) *Conversely, if $S(E)$ is a domain and $B(E) = C(E)$ then E satisfies $\mathcal{F}_2$.*

Proof. We may assume that R is a local ring. On account of $\mathcal{F}_2$, $\dim S(E) = \dim R + \text{rank }(E)$, and therefore $\dim S(E) = \dim D(E) = \dim C(E)$.

Let f be a homogeneous element of $B(E)$. The set

$$I = \{r \in R \mid r \cdot f \in C(E)\}$$

is an ideal of height at least 2. From Remark 1.3.5, we have height$(IS(E)) \geq 2$. If $I \neq R$ we shall find this impossible.

For simplicity we first argue the case $S(E) = D(E)$. Here we have

$$\text{height }(IC(E)) = \text{height }(IC(E) \bigcap D(E)) \geq \text{height }(ID(E)) \geq 2,$$

the equality on the left following from [44, Theorem 34.8]. As $C(E)$ is a Krull domain anf f lies in its field of fractions, this is impossible.

If $S(E)$ is not a domain, $D(E) = S(E)/P$, where P is a prime of height 0. In this case, height$((IS(E) + P)/P)$ is at least 2, and the argument applies.

For the converse, we verify $\mathcal{F}_2$ in terms of the local number of generators. Let P be the maximal ideal of R. We may assume height$(P) \geq 2$ and E is not free; we must show

$$v(E) \leq \text{height } (P) + \text{rank } (E) - 2 = \dim S(E) - 2.$$

The ideal I contains a regular sequence $\{a, b\}$ on R, which is also regular on $B(E)$, so we have height$(PB(E)) \geq 2$. By the result of Nagata,

$$\text{height } (PC(E)) = \text{height } (PC(E) \bigcap S(E)).$$

Because $PS(E)$ is a prime ideal of $S(E)$, $PC(E) \bigcap S(E) = PS(E)$, so height$(PS(E)) \geq 2$. This gives the required condition. $\square$

For the remainder of this section, R is either a polynomial ring over a field, or a geometric regular local ring. We shall now look at the Noetherianess of the factorial closure of a symmetric algebra $S(E)$. Since $B(E) = B(E^{**})$, for the bi-dual of E, we may henceforth assume that E is a reflexive module.

Here is a sketch of the method. Since the graded bi-dual of $S(E)$ or $D(E)$ coincide, we consider the embedding

$$D(E) \hookrightarrow B(E) = \sum_{j \geq 0} B_j.$$

Suppose the two algebras first differ in degree $r - 1$. We seek to determine the module C_r in the exact sequence

$$0 \longrightarrow D_r \longrightarrow B_r \dashrightarrow C_r \longrightarrow 0,$$

adding to $D(E)$ the necessary generators in degree r. The algebra obtained will be denoted by $B(r)$. This algebra is next checked for equality with $B(E)$; otherwise the preceding step is repeated for the next missing generators.

Here we shall only discuss the extreme case when $B(E)$ is obtainable from $D(E)$ by the addition of a single generator.

For simplicity sake, let R be a regular local ring, and let E be a reflexive module of projective dimension one, free on the punctured spectrum of R:

$$0 \longrightarrow R^m \xrightarrow{\varphi} R^n \longrightarrow E \longrightarrow 0.$$

We shall assume that $S(E)$ is a domain–equivalent here to the condition $\mathcal{F}_1$. If $\mathcal{F}_2$ does hold, $S(E)$ is a factorial domain. If $B(E) \neq S(E)$, we must have $m = d - 1, d = \dim R$.

The following result on strongly Cohen–Macaulay ideals is the necessary background for our development of the computation of $B(E)$.

Theorem 4.3.4 ([19, Theorem 2.6]) *Let R be a regular local ring and let I be a strongly Cohen-Macaulay ideal. If I satisfies the condition $\mathcal{F}_1$, then the associated graded ring*

$$gr_I(R) = \oplus_{i \geq 0} I^i/I^{i+1} \simeq S(I/I^2)$$

is a Gorenstein ring.

A consequence is that the extended Rees algebra of I

$$A = R[It, t^{-1}]$$

is a Gorenstein algebra as well. This formulation provides for a presentation for A once the ideal I has been given by its generators and relations. Specifically, suppose $I = (x_1, \ldots, x_n)$ has a presentation

$$R^m \xrightarrow{\varphi} R^n \longrightarrow I \longrightarrow 0, \ \varphi = (a_{ij}).$$

We obtain A as the quotient of the polynomial ring $R[T_1, \ldots, T_n, U]$ modulo the ideal J generated by the 1-forms in the T_i's

$$f_j = a_{1j}T_1 + \cdots + a_{nj}T_n, \ j = 1, \ldots, m,$$

together with the linear polynomials

$$UT_i - x_i, \ i = 1, \ldots, n.$$

Since the Krull dimension of A is $\dim R + 1$, J is an ideal of height n. Let us indicate the cases we shall make use of.

(a) If X is a generic $n - 1 \times n$ matrix (x_{ij}), and I is the ideal generated by the $n - 1$-sized minors of X, then

$$J = (UT_1 - D_1, \ldots, UT_n - D_n, \ \Sigma_i x_{ji}T_i, \ j = 1, \ldots, m).$$

By specializing $U = 0$, one obtains the ideal whose explicit resolution is given in [18].

(b) Let $X = (x_{ij})$ be the generic, skew–symmetric matrix of order $n + 1$, n even. Denote by I the ideal generated by the Pfaffians of X of order n. I is strongly Cohen-Macaulay and satisfies the condition on the local number of generators. J, in turn, is obtained as indicated above.

We require a strong assumption on the ideal $I_1(\varphi)$, that it be generated by a regular sequence. It is automatically satisfied if $S(E)$ is normal by Proposition 3.2.1, but it is present in other cases as well, e.g. in Example 3.2.2. With $I_1(\varphi) = (x_1, \ldots, x_d)$, as in Proposition 3.2.1, we write the presentation ideal $J(E)$ as

$$[x_1, \ldots, x_d] \cdot J(\varphi),$$

where $J(\varphi) = (b_{ij})$ is a $d - 1 \times d$ matrix of linear forms in the T_i-variables. Let D be the ideal generated by the $(d - 1)$–sized minors of $J(\varphi)$. $D = (D_1, \ldots, D_d)$ is a Cohen-Macaulay ideal of height two. An important role is that of the ideal D *evaluated at* $\mathbf{p}$, that is, taken mod $\mathbf{p}$; it will be denoted by Δ.

Note that normality requires $\Delta \neq 0$, while height $\Delta \leq 2$ by standard considerations. Δ can be interpreted as a kind of Jacobian ideal, and its height is independent of the choices made.

Theorem 4.3.5 ([55, Theorem 2.2]) *Let E be a module as above, and let $B(d-1)$ be the subalgebra of $B(E)$ obtained by adding to $S(E)$ the component of degree $d-1$ of $B(E)$. Then:*

(a) $B(d-1)$ *is a Gorenstein ring.*

(b) $B = B(d-1)$ *if and only if height $\Delta = 2$.*

Proof. We begin by showing that $S(E)$ and $B(E)$ first differ in degree $d-1$. Since E is free on the punctured spectrum and $S(E) \neq B(E)$, we must have in the resolution of E that $m = d - 1$. The projective resolution of a symmetric power $S_t(E)$ is then (*cf.* [62]):

$$0 \to \wedge^m R^m \otimes S_{t-m}(R^n) \to \wedge^{m-1} R^m \otimes S_{t-m+1}(R^n) \to \cdots \to S_t(R^n) \to 0.$$

Because this complex is exact and E is free outside of $\mathbf{p}$, we have that $S_t(E)$ is a reflexive module for $t \leq d-2$, but not reflexive outside this range. Furthermore, a direct calculation shows that

$$Ext^{d-1}(S_{d-1}(E), R) = R/I_1(\varphi).$$

Note that this measures the difference between B_{d-1} and S_{d-1}. Indeed, from the exact sequence

$$0 \longrightarrow S_{d-1} \longrightarrow B_{d-1} \longrightarrow C_{d-1} \longrightarrow 0$$

we obtain that

$$Ext^d(C_{d-1}, R) = Ext^{d-1}(S_{d-1}(E), R),$$

because B_{d-1} is a reflexive R-module. On the other hand, since the socle of $R/I_1(\varphi)$ is principal, we get, by duality, that $C_{d-1} = R/I_1(\varphi)$.

B_{d-1} is therefore obtainable from $S_{d-1}(E)$ by the addition of a single generator. We proceed to find this element. Denote by $D_1, \ldots, D_d$, the maximal minors of the matrix $J(\varphi)$. From the equations

$$f_j = 0, \ j = 1, \ldots, d-1$$

in $S(E)$, we obtain

$$x_i D_j \equiv x_j D_i.$$

Let u be the element D_1/x_1 of the field of fractions of $S(E)$. By [48], $u \in B$. We claim that $S(E)[u] = B(d-1)$, and that all of the asserted properties hold. First define the ideal $J(d-1)$ of the polynomial ring $P(d-1) = R[T_1, \ldots, T_n, U]$, generated by $J(E)$ and the polynomials

$$x_i U - D_i, \ 1 \leq i \leq d.$$

By hypothesis $J(E)$ is a prime ideal of height $d-1$, and as $\Delta \neq 0$, one of the forms D_i has unit content. Thus the ideal $J(d-1)$ has height at least d. It is then a proper specialization of the ideal (a) of Theorem 4.3.4 and is therefore a perfect Gorenstein ideal. Furthermore, since $\mathbf{p}P(d-1)$ is not an associated prime of $J(d-1)$, and E is free on the punctured spectrum of R, it follows easily that $J(d-1)$ is a prime ideal. This means that

$$P(d-1)/J(d-1) \simeq S(E)[u] \subseteq B(d-1).$$

We are now ready to prove the assertions. We begin by showing that u generates $B(d-1)$ over $S(E)$. Let v be the homogeneous generator of $B(d-1)$; we must have $u = bv + f$, where $b \in R$ and f is a form of $S(E)$ of degree $d-1$. We claim that b is a unit. Since the conductor of v, in degree 0, is $I_1(\varphi)$, the equations for v in degree two must be similar to those for u above, that is

$$x_i U - h_i = 0, \ 1 \le i \le d.$$

By assumption, one of the D_i does not have all of its coefficients in $\mathbf{p}$; assume D_1 is such a form. $x_1 bU - D_1 + x_1 f$ must be a linear combination of the linear equations for v:

$$x_1 bU - D_1 + x_1 f = \sum_i b_i (x_i U - h_i)$$

with

$$(b_1 - b)x_1 + \sum_{1 < i \le d} b_i x_i = 0.$$

But this is clearly impossible, unless b is a unit.

To prove (b) we use [57, Proposition 1.3.1]. As $B(d-1)$ is Cohen-Macaulay, we must have height $\mathbf{p}B(d-1) \ge 2$, a condition that is expressed by height $(\Delta) = 2$. $\square$

Example 4.3.6 Let us return to Example 3.2.2 and compute $B(E)$. The assumption on $I_1(\varphi)$ is realized since $\{a, b, c\}$ is a regular sequence.

Here

$$J(\varphi) = \begin{bmatrix} T_1 & T_2 \\ T_2 & T_3 \\ T_4 & T_5 \end{bmatrix}.$$

Because height$(\Delta) = 2$, $B(E) = B(2)$.

For another group of examples, let E be a reflexive module such that its symmetric algebra is an almost complete intersection. This means, according to Theorem 2.4.4, that E has a projective resolution

$$0 \longrightarrow R \overset{\psi}{\longrightarrow} R^m \overset{\varphi}{\longrightarrow} R^n \longrightarrow E \longrightarrow 0.$$

Whether $S(E)$ is an integral domain depends on the Koszul homology of the ideal $I = I_1(\psi)$.

We consider modules that are reflexive and free on the punctured spectrum of a regular local ring R. It implies that (with $\mathcal{F}_1$) $m = \dim R$. According to [54] this has the following consequence: If m is even then $S_{m/2}(E)$ is not a torsion-free R-module.

We focus on the case m odd. An application of the complexes of [62]—or through the approximation complexes—shows that $S_t(E)$ is reflexive for $t \le (m-3)/2$ and torsion-free but not reflexive if $t = r = (m-1)/2$. We seek the equations for $B(r)$.

Lemma 4.3.7 $B(E)_r/S_r(E)$ *is a cyclic R-module.*

Proof. As in the proof of Theorem 4.3.5 we consider the exact sequence

$$0 \longrightarrow S_r(E) \longrightarrow B(E)_r \longrightarrow C_r \longrightarrow 0,$$

and compute $Ext^{m-1}(S_r(E), R) = Ext^m(C_r, R)$. A direct application of the complexes of [62] shows that this module is isomorphic to $R/I_1(\psi)$. Because $I_1(\psi)$ is generated by a system of parameters, by duality it follows that C_r is cyclic. $\square$

To get the equations for $B(r)$, first observe that the rows of the matrix φ are syzygies of the system of parameters $I_1(\psi) = (x)$. In particular $I_1(\varphi)$ is contained in (x). Denote by P the polynomial ring $R[T_1, \dots, T_n]$. If $S(E)$ is a normal domain, we can write its defining ideal

$$J(E) = (f) = (x) \cdot J(\varphi).$$

Consider the exact sequence:

$$P^m \xrightarrow{J(\varphi)} P^m \longrightarrow C \longrightarrow 0.$$

Assume that $J(\varphi)$ has rank $m - 1$; thus ker $J(\varphi)$ is cyclic, generated, say, by the vector of forms $(g) = (g_1, \dots, g_m)$. Furthermore the associated prime ideals of C as an R-module are trivial–as E is free in codimension at most two–and therefore C is actually an ideal of P. It follows easily that C is a Gorenstein ideal and thus isomorphic to G, the ideal generated by the entries of (g).

Theorem 4.3.8 *Let E be an R-module that is reflexive and free on the punctured spectrum of R, whose symmetric algebra is a normal domain.*

(a) *$B(r)$ is a Gorenstein ring singly generated over $S(E)$.*

(b) *Denote by $G(0)$ the ideal G evaluated at the origin. Then $B(r) = B$ if and only if height $G(0) \geq 2$.*

Proof. Consider the equations

$$(x) \cdot J(\varphi) = (f)$$
$$(g) \cdot J(\varphi) = 0$$

Reading them mod (f), that is, in $S(E)$, when $J(\varphi)$ still has rank $m - 1$ by the normality hypothesis, we get an element h in the field of fractions of $S(E)$, such that $h \cdot (x) \equiv (g)$. It follows from [48] that h actually lies in B.

(a) Let $J(r)$ be the ideal of the polynomial ring $P[U]$

$$((f), (x) \cdot U - (g)).$$

$J(r)$ is by Theorem 4.3.4 a Gorenstein prime ideal of height m.

(b) Because B is the ideal transform of $B(r)$ with respect to the ideal (x), these rings are equal if and only if the grade of $(x) \cdot B(r)$ is at least two. Since $B(r)$ is Cohen–Macaulay this translates into the condition above. $\square$

Remark 4.3.9 Of course, if $S(E)$ is normal, the rank of $J(\varphi)$ is at least $m - 1$. It is likely to be $m - 1$ in all cases $S(E)$ is a domain. If $S(E)$ is not a domain, then the rank of $J(\varphi)$ may be higher, as in one of the examples below.

Example 4.3.10 In Example 2.4.5, an application of the *Macaulay* program yielded that $J(E)$ is a Cohen–Macaulay prime ideal. The ideal $G(0)$ is given by the Pfaffians of its Jacobian matrix $J(\varphi)$, computed earlier. It has height 3, so $B(E) = B(2)$.

For $n = 6$ the module is defined by the matrix

$$
\begin{bmatrix}
-x_2 & x_1 & 0 & 0 & 0 & 0 \\
-x_3 & 0 & x_1 & 0 & 0 & 0 \\
-x_4 & x_3 & -x_2 & x_1 & 0 & 0 \\
-x_5 & x_4 & 0 & -x_2 & x_1 & 0 \\
-x_6 & x_5 & -x_4 & x_3 & -x_2 & x_1 \\
0 & -x_6 & x_5 & 0 & -x_3 & x_2 \\
0 & 0 & -x_6 & x_5 & -x_4 & x_3 \\
0 & 0 & 0 & -x_6 & 0 & x_4 \\
0 & 0 & 0 & 0 & -x_6 & x_5
\end{bmatrix}.
$$

As remarked earlier, $S(E)$ cannot be an integral domain. A computation, showed that $D(E)$ is defined by the linear forms arising out of this matrix plus the polynomial:

$$T_5^3 + 2T_4T_5T_6 + T_3T_6^2 + T_4^2T_7 - 2T_3T_5T_7 - T_2T_6T_7 + T_1T_7^2 - T_3T_4T_8 + T_2T_5T_8 - T_1T_6T_8 + T_3^2T_9 - T_2T_4T_9 - T_1T_5T_9.$$

It was further verified that $B(E) = D(E)$, and that it is a Cohen-Macaulay ring.

The formalism used in the normality criterion may be used (*cf.* [22]) to view from another angle the symmetric algebra of a module with a linear presentation.

Given an R–module E with a presentation

$$R^m \xrightarrow{\varphi} R^n \longrightarrow E \longrightarrow 0, \quad \varphi = (a_{ij}),$$

the ideal of definition of its symmetric algebra can be written as a matrix product

$$J = [f_1, \ldots, f_m] = \mathbf{T} \cdot \varphi, \quad \mathbf{T} = [T_1, \ldots, T_m],$$

in an essentially unique manner.

Assume that R is a polynomial ring $k[\mathbf{x}] = k[x_1, \ldots, x_d]$ over some base ring k, and that the entries of φ are k–linear forms in the variables x_i. As above we can write the ideal J as

$$J = \mathbf{x} \cdot B, \quad B = (b_{\ell j}),$$

where B is an $d \times m$ matrix of k–linear forms in the variables T_ℓ.

Let F be the $k[\mathbf{T}] = k[T_1, \ldots, T_n]$–module defined by the matrix B:

$$k[\mathbf{T}]^m \xrightarrow{B} k[\mathbf{T}]^d \longrightarrow F \longrightarrow 0.$$

One has the identification

Proposition 4.3.11 $S = S_{k[\mathbf{x}]}(E) = S_{k[\mathbf{T}]}(F)$.

This permits toggling between representations. Thus a question over a d–dimensional ring turns into the same question on a d–generated module over the other ring. For instance, if S is an integral domain (with $k =$ field), its Krull dimension is $d+r_{\mathbf{X}} = n+r_{\mathbf{T}}$, where $r_{\mathbf{X}}$ and $r_{\mathbf{T}}$ are the ranks of E and F respectively.

For an application, consider Example 2.4.5 with $n = 5$. The module E has rank 3, while F has rank 1. The matrix B is skew symmetric and it is easy to see that F can be identified to the ideal generated by the Pfaffians of B. It will follow that $S(E)$ is Cohen–Macaulay. [1]

4.4   Symbolic power algebras

We touch briefly on a question similar to finding the factorial closure of a symmetric algebra. It concerns the computation of symbolic blow-ups. The main result here is a formula describing the second symbolic power of almost complete intersections.

Let R be a regular local ring of dimension d and let I be a prime ideal of height g. Denote by $I^{(2)}$ the I-primary component of I^2. One way to express the equality–or difference–between I^2 and $I^{(2)}$ is through the module I/I^2: $I^2 = I^{(2)}$ if and only if I/I^2 is a torsion-free R/I-module. Let us seek to determine the torsion of this module.

Let $I = (x_1, \ldots, x_n)$; there exists an exact sequence

$$H_1 \xrightarrow{\psi} (R/I)^n \longrightarrow I/I^2 \longrightarrow 0$$

where H_1 is the first Koszul homology module of I. Let us assume that H_1 is a Cohen-Macaulay module–which will make the sequence exact on the left—and further that $\dim(R/I) = 1$. The long exact sequence of the functor $Ext(-, R)$ yields ($S = R/I$):

$$Ext^{d-1}(S^n, R) \longrightarrow Ext^{d-1}(H_1, R) \longrightarrow Ext^d(I/I^2, R) \longrightarrow 0.$$

On the other hand, the sequence

$$0 \longrightarrow I^{(2)}/I^2 \longrightarrow I/I^2 \longrightarrow I/I^{(2)} \longrightarrow 0$$

gives rise to an isomorphism $Ext^d(I^{(2)}/I^2, R) = Ext^d(I/I^2, R)$, since $I/I^{(2)}$ is a torsion-free S-module and therefore Cohen-Macaulay by the dimension condition. If we now let W denote the canonical module of S, $W = Ext^{d-1}(S, R)$, and use the duality in the Koszul homology ([20]), we get the exact sequence

$$W^n = Hom(S^n, W) \xrightarrow{\varphi^\bullet} Hom(H_1, W) \longrightarrow Ext^d(I^{(2)}/I^2, R) \longrightarrow 0.$$

Note that if $[z = \sum z_i e_i]$ is an element of H_1, $\varphi([z]) = \sum \overline{z_i} e_i \in S^n$. From this we can read the image of φ^*: For any $\lambda_w \in W^n$, $\lambda_w(s_1, \ldots, s_d) = \sum s_i w_i$, $\varphi^*(\lambda_w([z])) = \sum z_i w_i$.

Proposition 4.4.1 *Let I be an almost complete intersection of height $d - 1$. Then $I^{(2)}/I^2 = Ext^d(R/I_1(\varphi), R)$, where $I_1(\varphi)$ denotes the ideal generated by the entries of the first order syzygies of I.*

---

[1]Added in proof: It can be shown that the factorial closures of modules such as Example refve are always Noetherian.

Proof. Pick a generating set for I such that any $d-1$ elements in it form a regular sequence. To find the image of φ^* in $Hom(H_1 = W, W) = S$, it suffices to observe that for any two cycles $\alpha = \sum a_i e_i$ and $\beta = \sum b_i e_i$ we have $a_i \beta = b_i \alpha$ in S^n. It follows that the image of φ^* is precisely $I_1(\varphi)$.

Corollary 4.4.2 *Let I be a prime ideal of a regular local ring $(R, \mathbf{m})$. If I is a normal, almost complete intersection of dimension one, then $I^{(2)}/I^2$ is cyclic. Furthermore if $I \subset \mathbf{m}^2$, then the Cohen-Macaulay type of I is at least $\dim R - 1$.*

Proof. Because the symmetric and Rees algebras of I coincide [20] and it is Cohen-Macaulay, by the normality criterion of Proposition 3.2.1 we not only must have $I_1(\varphi) = (y_1, \ldots, y_{d-1}, y_d^a)$ for some regular system of parameters (y), which proves the first assertion, but also the rank of the Jacobian matrix at the origin must be $d-1$. Picking a generating set for I such that $\{x_1, \ldots, x_{d-1}\}$ is a regular set, as $x_i \in \mathbf{m}^2$, this implies that there must be at least $d-1$ syzygies that contribute to the non-vanishing of the Jacobian ideal. $\square$

At the moment we are puzzled as to where to find the generator of $I^{(2)}/I^2$ as a determinantal element of the syzygies of I, in the manner of Theorem 4.3.5.

4.5 Roberts construction

We shall now give Roberts' example [47] of a module whose factorial closure is not Noetherian. Actually, his aim was twofold: (i) To find a counterexample to Hilbert's 14th Problem, and (ii) to construct a prime ideal in a power series ring whose symbolic blow-up is not Noetherian. (His earlier example [46] of the latter was not analytically irreducible.)

There are two parts to his construction. First, establishing a setting whereby the construction permits deciding several features of the factorial closure; then, a delicate analysis of a family of examples.

Let R be a regular local ring (*resp.* a polynomial ring over a field), and let E be a finitely generated module (*resp.* a graded R-module). Recall that in studying the factorial closure of $S(E)$, we may assume that E is a reflexive module. This allows for an exact sequence

$$0 \longrightarrow E \overset{\psi}{\longrightarrow} F \overset{\varphi}{\longrightarrow} G \longrightarrow 0,$$

where F is a free module and G is a torsion-free module.

If $T_1, \ldots, T_n$ is a basis of F, E is generated by some linear forms

$$f_j = \sum_{i=1}^{n} a_{ij} T_i, \ j = 1, \ldots, m.$$

$D(E)$ is the subring of $R[T_1, \ldots, T_n]$ generated by $f_1, \ldots, f_m$ over the base ring R. At this point we can assume that $E \subset mF$, $m = $ maximal ideal of R. When E is a free module this is exactly $B(E)$. Let I be any height two ideal in the ideal defining the free locus of E.

Lemma 4.5.1 $B(E)$ *consists of the elements of* $R[T_1, \ldots, T_n]$ *conducted into* $D(E)$ *by some power of* I.

This follows from the earlier discussion.

However explicit, this does not provide the setting to carry out computations. The other leg of the construction is the so-called downgrading homomorphism in the symmetric algebra of a module.

Let $\varphi : F \longrightarrow R$ be homomorphism of an R–module. There is a R–module endomorphism of $S(F)$ that extends φ. It is defined as follows. Set $\varphi_0 = 0$ and $\varphi_1 = \varphi$. Then $\varphi_n : S_n(F) \longrightarrow S_{n-1}(F)$ is given by

$$\varphi_n(\sum e_1 \cdots e_n) = \sum \varphi(e_i)(e_1 \cdots \hat{e}_i \cdots e_n).$$

Proposition 4.5.2 *Assume that the image of the mapping* $\varphi : F = R^r \longrightarrow R$ *is minimally generated by* r *elements. Then* $kernel(\varphi_n) \subset mS_n(F)$.

Actually this is somewhat misleading. If R contains a field of characteristic 0, the construction stands as above. In general however $S(F)$ has to be replaced by the divided powers algebra.

Proof. Once a basis $\{e_1, \ldots, e_r\}$ is chosen, $S_n(F)$ is freely generated by the monomials $e_1^{a_1} \cdots e_r^{a_r}$, $\sum a_i = n$. If such monomial occurs, with coefficient 1, in an element of the kernel of φ_n, then $a_1 \varphi(e_1) e_1^{a_1-1} \cdots e_r^{a_r}$ (say $a_1 \geq 1$) would be a combination of elements all with coefficients in $\varphi(Re_2 + \cdots + Re_r)$, contradicting the minimality hypothesis. $\square$

One uses this in conjunction with the previous lemma. That is, assume that we have an exact sequence

$$0 \longrightarrow E \longrightarrow F = R^r \longrightarrow I \longrightarrow 0,$$

with I an ideal minimally generated by r elements. There is a surjective homomorphism from the symmetric algebra of F onto the Rees algebra of I:

$$S(F) \xrightarrow{\alpha} S(I) \xrightarrow{\beta} \mathcal{R}(I)$$

where α is induced by φ and β is the natural mapping.

The kernel J of α is generated by the 1-forms that define E; that of $\beta \circ \alpha$, J_∞, will be larger, if I is not of linear type.

Corollary 4.5.3 *In the situation above* $(K = \text{field of quotients of } R)$:

$$\begin{cases} J_\infty = \sum_{i \geq 1} kernel(\varphi_i \circ \cdots \circ \varphi_1) \\ B_n(E) = kernel(\varphi_n) = (J_\infty)_n \bigcap (S(E) \otimes K) \end{cases}$$

This implies that the components $B_n(E)$ lie in $mS_n(F)$, so that if $B(E)$ is finitely generated then for all large n, $B_n(E)$ will be contained in $m^\ell S_n(F)$ for ℓ large as well.

Here is the critical result of Roberts:

Theorem 4.5.4 *Let* I *be the ideal of* $k[x, y, z]$ *generated by* x^{t+1}, y^{t+1}, z^{t+1}, $x^t y^t z^t$, *for some integer* $t \geq 2$. *For each integer* n *there exists an element in* $B_n(E)$ *whose coefficient of* T_4^n *is* x.

References

[1] M. Auslander, Modules over unramified regular local rings, Illinois J. Math. **5** (1961), 631–647.

[2] M. Auslander and O. Goldman, Maximal orders, Trans. Amer. Math. Soc. **97** (1960), 1–24.

[3] L. Avramov, Complete intersections and symmetric algebras, J. Algebra **73** (1980), 249–280.

[4] J. Barshay, Graded algebras of powers of ideals generated by A–sequences, J. Algebra **25** (1973), 90–99.

[5] D. Bayer and M. Stillman, *Macaulay*, A computer algebra system for computing in algebraic geometry and commutative algebra, 1988.

[6] J. Brennan, M. Vaz Pinto and W. V. Vasconcelos, The Jacobian module of a Lie algebra, Trans. Amer. Math. Soc., to appear.

[7] W. Bruns, Additions to the theory of algebras with straightening law, in *Commutative Algebra* (M. Hochster, C. Huneke and J. D. Sally, Eds.), MSRI Publications **15**, Springer–Verlag, New York, 1989, 111–138.

[8] D. Buchsbaum and D. Eisenbud, What makes a complex exact?, J. Algebra **25** (1973), 259–268.

[9] D. Costa, L. Gallardo and J. Querré, On the distribution of prime elements in polynomial Krull domains, Proc. Amer. Math. Soc. **87** (1983), 41–43.

[10] D.L. Costa and J.L. Johnson, Inert extensions for Krull domains, Proc. Amer. Math. Soc. **59** (1976), 189–194.

[11] J. Eagon and D. G. Northcott, Ideals defined by matrices and a certain complex associated with them, Proc. Royal Soc. **269** (1962), 188–204.

[12] E. G. Evans and P. Griffith, The syzygy problem, Annals of Math. **114** (1981), 323–333.

[13] H. Fitting, Die Determinantenideale Moduls, Jahresbericht DMV **46** (1936), 192–228.

[14] O. Forster, Über die Anzahl der Erzeugenden eines Ideals in einem Noetherschen Ring, Math. Z. **84** (1964), 80–87.

[15] F. R. Gantmacher, *Applications of the Theory of Matrices*, Interscience, New York, 1959.

[16] M. Gerstenhaber, On dominance and varieties of commuting matrices, Annals of Math. **73** (1961), 324–348.

[17] R. Gilmer, *Multiplicative Ideal Theory*, M. Dekker, New York, 1972.

[18] J. Herzog, Certain complexes associated to a sequence and a matrix, Manuscripta Math. **12** (1974), 217–247.

[19] J. Herzog, A. Simis and W. V. Vasconcelos, Approximation complexes of blowing–up rings, J. Algebra **74** (1982), 466–493.

[20] J. Herzog, A. Simis and W. V. Vasconcelos, Koszul homology and blowing–up rings, Proc. Trento Commutative Algebra Conf., Lectures Notes in Pure and Applied Math., vol. **84**, Dekker, New York, 1983, 79–169.

[21] J. Herzog, A. Simis and W. V. Vasconcelos, On the arithmetic and homology of algebras of linear type, Trans. Amer. Math. Soc. **283** (1984), 661–683.

[22] J. Herzog, A. Simis and W. V. Vasconcelos, Arithmetic of normal Rees algebras, Preprint, 1988.

[23] J. Herzog, W. V. Vasconcelos and R. Villarreal, Ideals with sliding depth, Nagoya Math. J. **99** (1985), 159–172.

[24] M. Hochster, The Zariski–Lipman conjecture in the graded case, J. Algebra **47** (1977), 411–424.

[25] J. E. Humphreys, *Introduction to Lie Algebras and Representation Theory*, Springer–Verlag, Berlin, 1972.

[26] C. Huneke, On the symmetric algebra of a module, J. Algebra **69** (1981), 113–119.

[27] C. Huneke, Determinantal ideals of linear type, Arch. Math. **47** (1986), 324–329.

[28] C. Huneke, On the symmetric and Rees algebras of an ideal generated by a d–sequence, J. Algebra **62** (1980), 268–275.

[29] C. Huneke, Linkage and Koszul homology of ideals, Amer. J. Math. **104** (1982), 1043–1062.

[30] C. Huneke, Strongly Cohen–Macaulay schemes and residual intersections, Trans. Amer. Math. Soc. **277** (1983), 739–763.

[31] C. Huneke and M. E. Rossi, The dimension and components of symmetric algebras, J. Algebra **98** (1986), 200–210.

[32] C. Huneke and B. Ulrich, Residual intersections, J. reine angew. Math. **390** (1988), 1–20.

[33] T. Jòsefiak, Ideals generated by the minors of a symmetric matrix, Comment. Math. Helvetici **53** (1978), 595–607.

[34] B. Kostant, Lie group representations on polynomial rings, Amer. J. Math. **85** (1963), 327–404.

[35] B. V. Kotsev, Determinantal ideals of linear type of a generic symmetric matrix, J. Algebra, to appear.

[36] L. Lebelt, Freie Auflösungen äußerer Potenzen, Manuscripta Math. **21** (1977), 341–355.

[37] S. Lichtenbaum, On the vanishing of Tor in regular local rings, Illinois J. Math. **10** (1966), 220–226.

[38] J. Lipman, Free derivation modules on algebraic varieties, American J. Math. **87** (1965), 874–898.

[39] R. MacRae, On an application of the Fitting invariants, J. Algebra **2** (1965), 153–169.

[40] H. Matsumura, *Commutative Algebra*, Benjamin/Cummings, Reading, Massachusetts, 1980.

[41] A. Micali, Sur les algèbres universalles, Annales Inst. Fourier **14** (1964), 33–88.

[42] A. Micali, P. Salmon and P. Samuel, Integrité et factorialité des algèbres symétriques, Atas do IV Colóquio Brasileiro de Matemática, SBM, (1965), 61–76.

[43] T. Motzkin and O. Taussky, Pairs of matrices with property L II, Trans. Amer. Math. Soc. **80** (1955), 387–401.

[44] M. Nagata, *Local Rings*, Interscience, New York, 1962.

[45] R. W. Richardson, Commuting varieties of semisimple Lie algebras and algebraic groups, Compositio Math. **38** (1979), 311–327.

[46] P. Roberts, A prime ideal in a polynomial ring whose symbolic blow-up is not Noetherian, Proc. Amer. Math. Soc. **94** (1985), 589–592.

[47] P. Roberts, An infinitely generated symbolic blow–up in a power series ring and a new counterexample to Hilbert's Fourteenth Problem, J. Algebra, to appear.

[48] P. Samuel, Anneaux gradués factoriels et modules réflexifs, Bull. Soc. Math. France **92** (1964), 237–249.

[49] A. Simis, *Selected Topics in Commutative Algebra*, Lecture Notes, IX Escuela Latinoamericana de Matematica, Santiago, Chile, 1988.

[50] A. Simis and W. V. Vasconcelos, The syzygies of the conormal module, Amer. J. Math. **103** (1981), 203–224.

[51] A. Simis and W. V. Vasconcelos, On the dimension and integrality of symmetric algebras, Math. Z. **177** (1981), 341–358.

[52] A. Simis and W. V. Vasconcelos, The Krull dimension and integrality of symmetric algebras, Manuscripta Math. **61** (1988), 63–78.

[53] G. Valla, On the symmetric and Rees algebras of an ideal, Manuscripta Math. **30** (1980), 239–255.

[54] W. V. Vasconcelos, On linear complete intersections, J. Algebra **111** (1987), 306–315.

[55] W. V. Vasconcelos, Symmetric algebras and factoriality, in *Commutative Algebra* (M. Hochster, C. Huneke and J. D. Sally, Eds.), MSRI Publications **15**, Springer-Verlag, New York, 1989, 467–496.

[56] W. V. Vasconcelos, The complete intersection locus of certain ideals, J. Pure and Applied Algebra **38** (1985), 367–378.

[57] W. V. Vasconcelos, On the structure of certain ideal transforms, Math. Z. **198** (1988), 435–448.

[58] W. V. Vasconcelos, Modules of differentials of symmetric algebras, Arch. Math., to appear.

[59] U. Vetter, Zu einem Satz von G. Trautmann über den Rang gewisser kohärenter analytischer Moduln, Arch. Math. **24** (1973), 158–161.

[60] R. Villarreal, Rees algebras and Koszul homology, J. Algebra **119** (1988), 83–104.

[61] R. Villarreal, Cohen–Macaulay graphs, Preprint, 1988.

[62] J. Weyman, Resolutions of the exterior and symmetric powers of a module, J. Algebra **58** (1979), 333–341.

Department of Mathematics, Rutgers University, New Brunswick, New Jersey 08903, USA

Vol. 1259: F. Cano Torres, Desingularization Strategies for Three-Dimensional Vector Fields. IX, 189 pages. 1987.

Vol. 1260: N.H. Pavel, Nonlinear Evolution Operators and Semigroups. VI, 285 pages. 1987.

Vol. 1261: H. Abels, Finite Presentability of S-Arithmetic Groups. Compact Presentability of Solvable Groups. VI, 178 pages. 1987.

Vol. 1262: E. Hlawka (Hrsg.), Zahlentheoretische Analysis II. Seminar, 1984–86. V, 158 Seiten. 1987.

Vol. 1263: V.L. Hansen (Ed.), Differential Geometry. Proceedings, 1985. XI, 288 pages. 1987.

Vol. 1264: Wu Wen-tsün, Rational Homotopy Type. VIII, 219 pages. 1987.

Vol. 1265: W. Van Assche, Asymptotics for Orthogonal Polynomials. VI, 201 pages. 1987.

Vol. 1266: F. Ghione, C. Peskine, E. Sernesi (Eds.), Space Curves. Proceedings, 1985. VI, 272 pages. 1987.

Vol. 1267: J. Lindenstrauss, V.D. Milman (Eds.), Geometrical Aspects of Functional Analysis. Seminar. VII, 212 pages. 1987.

Vol. 1268: S.G. Krantz (Ed.), Complex Analysis. Seminar, 1986. VII, 195 pages. 1987.

Vol. 1269: M. Shiota, Nash Manifolds. VI, 223 pages. 1987.

Vol. 1270: C. Carasso, P.-A. Raviart, D. Serre (Eds.), Nonlinear Hyperbolic Problems. Proceedings, 1986. XV, 341 pages. 1987.

Vol. 1271: A.M. Cohen, W.H. Hesselink, W.L.J. van der Kallen, J.R. Strooker (Eds.), Algebraic Groups Utrecht 1986. Proceedings. XII, 284 pages. 1987.

Vol. 1272: M.S. Livšic, L.L. Waksman, Commuting Nonselfadjoint Operators in Hilbert Space. III, 115 pages. 1987.

Vol. 1273: G.-M. Greuel, G. Trautmann (Eds.), Singularities, Representation of Algebras, and Vector Bundles. Proceedings, 1985. XIV, 383 pages. 1987.

Vol. 1274: N. C. Phillips, Equivariant K-Theory and Freeness of Group Actions on C*-Algebras. VIII, 371 pages. 1987.

Vol. 1275: C.A. Berenstein (Ed.), Complex Analysis I. Proceedings, 1985–86. XV, 331 pages. 1987.

Vol. 1276: C.A. Berenstein (Ed.), Complex Analysis II. Proceedings, 1985–86. IX, 320 pages. 1987.

Vol. 1277: C.A. Berenstein (Ed.), Complex Analysis III. Proceedings, 1985–86. X, 350 pages. 1987.

Vol. 1278: S.S. Koh (Ed.), Invariant Theory. Proceedings, 1985. V, 102 pages. 1987.

Vol. 1279: D. Ieşan, Saint-Venant's Problem. VIII, 162 Seiten. 1987.

Vol. 1280: E. Neher, Jordan Triple Systems by the Grid Approach. XII, 193 pages. 1987.

Vol. 1281: O.H. Kegel, F. Menegazzo, G. Zacher (Eds.), Group Theory. Proceedings, 1986. VII, 179 pages. 1987.

Vol. 1282: D.E. Handelman, Positive Polynomials, Convex Integral Polytopes, and a Random Walk Problem. XI, 136 pages. 1987.

Vol. 1283: S. Mardešić, J. Segal (Eds.), Geometric Topology and Shape Theory. Proceedings, 1986. V, 261 pages. 1987.

Vol. 1284: B.H. Matzat, Konstruktive Galoistheorie. X, 286 pages. 1987.

Vol. 1285: I.W. Knowles, Y. Saitō (Eds.), Differential Equations and Mathematical Physics. Proceedings, 1986. XVI, 499 pages. 1987.

Vol. 1286: H.R. Miller, D.C. Ravenel (Eds.), Algebraic Topology. Proceedings, 1986. VII, 341 pages. 1987.

Vol. 1287: E.B. Saff (Ed.), Approximation Theory, Tampa. Proceedings, 1985–1986. V, 228 pages. 1987.

Vol. 1288: Yu. L. Rodin, Generalized Analytic Functions on Riemann Surfaces. V, 128 pages, 1987.

Vol. 1289: Yu. I. Manin (Ed.), K-Theory, Arithmetic and Geometry. Seminar, 1984–1986. V, 399 pages. 1987.

Vol. 1290: G. Wüstholz (Ed.), Diophantine Approximation and Transcendence Theory. Seminar, 1985. V, 243 pages. 1987.

Vol. 1291: C. Mœglin, M.-F. Vignéras, J.-L. Waldspurger, Correspondances de Howe sur un Corps p-adique. VII, 163 pages. 1987

Vol. 1292: J.T. Baldwin (Ed.), Classification Theory. Proceedings, 1985. VI, 500 pages. 1987.

Vol. 1293: W. Ebeling, The Monodromy Groups of Isolated Singularities of Complete Intersections. XIV, 153 pages. 1987.

Vol. 1294: M. Queffélec, Substitution Dynamical Systems – Spectral Analysis. XIII, 240 pages. 1987.

Vol. 1295: P. Lelong, P. Dolbeault, H. Skoda (Réd.), Séminaire d'Analyse P. Lelong – P. Dolbeault – H. Skoda. Seminar, 1985/1986. VII, 283 pages. 1987.

Vol. 1296: M.-P. Malliavin (Ed.), Séminaire d'Algèbre Paul Dubreil et Marie-Paule Malliavin. Proceedings, 1986. IV, 324 pages. 1987.

Vol. 1297: Zhu Y.-l., Guo B.-y. (Eds.), Numerical Methods for Partial Differential Equations. Proceedings. XI, 244 pages. 1987.

Vol. 1298: J. Aguadé, R. Kane (Eds.), Algebraic Topology, Barcelona 1986. Proceedings. X, 255 pages. 1987.

Vol. 1299: S. Watanabe, Yu.V. Prokhorov (Eds.), Probability Theory and Mathematical Statistics. Proceedings, 1986. VIII, 589 pages. 1988.

Vol. 1300: G.B. Seligman, Constructions of Lie Algebras and their Modules. VI, 190 pages. 1988.

Vol. 1301: N. Schappacher, Periods of Hecke Characters. XV, 160 pages. 1988.

Vol. 1302: M. Cwikel, J. Peetre, Y. Sagher, H. Wallin (Eds.), Function Spaces and Applications. Proceedings, 1986. VI, 445 pages. 1988.

Vol. 1303: L. Accardi, W. von Waldenfels (Eds.), Quantum Probability and Applications III. Proceedings, 1987. VI, 373 pages. 1988.

Vol. 1304: F.Q. Gouvêa, Arithmetic of p-adic Modular Forms. VIII, 121 pages. 1988.

Vol. 1305: D.S. Lubinsky, E.B. Saff, Strong Asymptotics for Extremal Polynomials Associated with Weights on IR. VII, 153 pages. 1988.

Vol. 1306: S.S. Chern (Ed.), Partial Differential Equations. Proceedings, 1986. VI, 294 pages. 1988.

Vol. 1307: T. Murai, A Real Variable Method for the Cauchy Transform, and Analytic Capacity. VIII, 133 pages. 1988.

Vol. 1308: P. Imkeller, Two-Parameter Martingales and Their Quadratic Variation. IV, 177 pages. 1988.

Vol. 1309: B. Fiedler, Global Bifurcation of Periodic Solutions with Symmetry. VIII, 144 pages. 1988.

Vol. 1310: O.A. Laudal, G. Pfister, Local Moduli and Singularities. V, 117 pages. 1988.

Vol. 1311: A. Holme, R. Speiser (Eds.), Algebraic Geometry, Sundance 1986. Proceedings. VI, 320 pages. 1988.

Vol. 1312: N.A. Shirokov, Analytic Functions Smooth up to the Boundary. III, 213 pages. 1988.

Vol. 1313: F. Colonius, Optimal Periodic Control. VI, 177 pages. 1988.

Vol. 1314: A. Futaki, Kähler-Einstein Metrics and Integral Invariants. IV, 140 pages. 1988.

Vol. 1315: R.A. McCoy, I. Ntantu, Topological Properties of Spaces of Continuous Functions. IV, 124 pages. 1988.

Vol. 1316: H. Korezlioglu, A.S. Ustunel (Eds.), Stochastic Analysis and Related Topics. Proceedings, 1986. V, 371 pages. 1988.

Vol. 1317: J. Lindenstrauss, V.D. Milman (Eds.), Geometric Aspects of Functional Analysis. Seminar, 1986–87. VII, 289 pages. 1988.

Vol. 1318: Y. Felix (Ed.), Algebraic Topology – Rational Homotopy. Proceedings, 1986. VIII, 245 pages. 1988

Vol. 1319: M. Vuorinen, Conformal Geometry and Quasiregular Mappings. XIX, 209 pages. 1988.

Vol. 1320: H. Jürgensen, G. Lallement, H.J. Weinert (Eds.), Semigroups, Theory and Applications. Proceedings, 1986. X, 416 pages. 1988.

Vol. 1321: J. Azéma, P.A. Meyer, M. Yor (Eds.), Séminaire de Probabilités XXII. Proceedings. IV, 600 pages. 1988.

Vol. 1322: M. Métivier, S. Watanabe (Eds.), Stochastic Analysis. Proceedings, 1987. VII, 197 pages. 1988.

Vol. 1323: D.R. Anderson, H.J. Munkholm, Boundedly Controlled Topology. XII, 309 pages. 1988.

Vol. 1324: F. Cardoso, D.G. de Figueiredo, R. Iório, O. Lopes (Eds.), Partial Differential Equations. Proceedings, 1986. VIII, 433 pages. 1988.

Vol. 1325: A. Truman, I.M. Davies (Eds.), Stochastic Mechanics and Stochastic Processes. Proceedings, 1986. V, 220 pages. 1988.

Vol. 1326: P.S. Landweber (Ed.), Elliptic Curves and Modular Forms in Algebraic Topology. Proceedings, 1986. V, 224 pages. 1988.

Vol. 1327: W. Bruns, U. Vetter, Determinantal Rings. VII,236 pages. 1988.

Vol. 1328: J.L. Bueso, P. Jara, B. Torrecillas (Eds.), Ring Theory. Proceedings, 1986. IX, 331 pages. 1988.

Vol. 1329: M. Alfaro, J.S. Dehesa, F.J. Marcellan, J.L. Rubio de Francia, J. Vinuesa (Eds.): Orthogonal Polynomials and their Applications. Proceedings, 1986. XV, 334 pages. 1988.

Vol. 1330: A. Ambrosetti, F. Gori, R. Lucchetti (Eds.), Mathematical Economics. Montecatini Terme 1986. Seminar. VII, 137 pages. 1988.

Vol. 1331: R. Bamón, R. Labarca, J. Palis Jr. (Eds.), Dynamical Systems, Valparaiso 1986. Proceedings. VI, 250 pages. 1988.

Vol. 1332: E. Odell, H. Rosenthal (Eds.), Functional Analysis. Proceedings, 1986–87. V, 202 pages. 1988.

Vol. 1333: A.S. Kechris, D.A. Martin, J.R. Steel (Eds.), Cabal Seminar 81–85. Proceedings, 1981–85. V, 224 pages. 1988.

Vol. 1334: Yu.G. Borisovich, Yu. E. Gliklikh (Eds.), Global Analysis – Studies and Applications III. V, 331 pages. 1988.

Vol. 1335: F. Guillén, V. Navarro Aznar, P. Pascual-Gainza, F. Puerta, Hyperrésolutions cubiques et descente cohomologique. XII, 192 pages. 1988.

Vol. 1336: B. Helffer, Semi-Classical Analysis for the Schrödinger Operator and Applications. V, 107 pages. 1988.

Vol. 1337: E. Sernesi (Ed.), Theory of Moduli. Seminar, 1985. VIII, 232 pages. 1988.

Vol. 1338: A.B. Mingarelli, S.G. Halvorsen, Non-Oscillation Domains of Differential Equations with Two Parameters. XI, 109 pages. 1988.

Vol. 1339: T. Sunada (Ed.), Geometry and Analysis of Manifolds. Procedings, 1987. IX, 277 pages. 1988.

Vol. 1340: S. Hildebrandt, D.S. Kinderlehrer, M. Miranda (Eds.), Calculus of Variations and Partial Differential Equations. Proceedings, 1986. IX, 301 pages. 1988.

Vol. 1341: M. Dauge, Elliptic Boundary Value Problems on Corner Domains. VIII, 259 pages. 1988.

Vol. 1342: J.C. Alexander (Ed.), Dynamical Systems. Proceedings, 1986–87. VIII, 726 pages. 1988.

Vol. 1343: H. Ulrich, Fixed Point Theory of Parametrized Equivariant Maps. VII, 147 pages. 1988.

Vol. 1344: J. Král, J. Lukeš, J. Netuka, J. Veselý (Eds.), Potential Theory – Surveys and Problems. Proceedings, 1987. VIII, 271 pages. 1988.

Vol. 1345: X. Gomez-Mont, J. Seade, A. Verjovski (Eds.), Holomorphic Dynamics. Proceedings, 1986. VII, 321 pages. 1988.

Vol. 1346: O. Ya. Viro (Ed.), Topology and Geometry – Rohlin Seminar. XI, 581 pages. 1988.

Vol. 1347: C. Preston, Iterates of Piecewise Monotone Mappings on an Interval. V, 166 pages. 1988.

Vol. 1348: F. Borceux (Ed.), Categorical Algebra and its Applications. Proceedings, 1987. VIII, 375 pages. 1988.

Vol. 1349: E. Novak, Deterministic and Stochastic Error Bounds in Numerical Analysis. V, 113 pages. 1988.

Vol. 1350: U. Koschorke (Ed.), Differential Topology. Proceedings 1987. VI, 269 pages. 1988.

Vol. 1351: I. Laine, S. Rickman, T. Sorvali, (Eds.), Complex Analysis Joensuu 1987. Proceedings. XV, 378 pages. 1988.

Vol. 1352: L.L. Avramov, K.B. Tchakerian (Eds.), Algebra – Some Current Trends. Proceedings, 1986. IX, 240 Seiten. 1988.

Vol. 1353: R.S. Palais, Ch.-l. Terng, Critical Point Theory and Submanifold Geometry. X, 272 pages. 1988.

Vol. 1354: A. Gómez, F. Guerra, M.A. Jiménez, G. López (Eds.), Approximation and Optimization. Proceedings, 1987. VI, 280 pages. 1988.

Vol. 1355: J. Bokowski, B. Sturmfels, Computational Synthetic Geometry. V, 168 pages. 1989.

Vol. 1356: H. Volkmer, Multiparameter Eigenvalue Problems and Expansion Theorems. VI, 157 pages. 1988.

Vol. 1357: S. Hildebrandt, R. Leis (Eds.), Partial Differential Equation and Calculus of Variations. VI, 423 pages. 1988.

Vol. 1358: D. Mumford, The Red Book of Varieties and Schemes. V, 309 pages. 1988.

Vol. 1359: P. Eymard, J.-P. Pier (Eds.), Harmonic Analysis. Proceedings, 1987. VIII, 287 pages. 1988.

Vol. 1360: G. Anderson, C. Greengard (Eds.), Vortex Methods. Proceedings, 1987. V, 141 pages. 1988.

Vol. 1361: T. tom Dieck (Ed.), Algebraic Topology and Transformation Groups. Proceedings, 1987. VI, 298 pages. 1988.

Vol. 1362: P. Diaconis, D. Elworthy, H. Föllmer, E. Nelson, G. C. Papanicolaou, S.R.S. Varadhan. École d'Été de Probabilités de Saint-Flour XV–XVII, 1985–87. Editor: P.L. Hennequin. V, 459 pages. 1988.

Vol. 1363: P.G. Casazza, T.J. Shura. Tsirelson's Space. VIII, 204 pages. 1988.

Vol. 1364: R.R. Phelps, Convex Functions, Monotone Operators and Differentiability. IX, 115 pages. 1989.

Vol. 1365: M. Giaquinta (Ed.), Topics in Calculus of Variations. Seminar, 1987. X, 196 pages. 1989.

Vol. 1366: N. Levitt, Grassmannians and Gauss Maps in PL-Topology. V, 203 pages. 1989.

Vol. 1367: M. Knebusch, Weakly Semialgebraic Spaces. XX, 376 pages. 1989.

Vol. 1368: R. Hübl, Traces of Differential Forms and Hochschild Homology. III, 111 pages. 1989.

Vol. 1369: B. Jiang, Ch.-K. Peng, Z. Hou (Eds.), Differential Geometry and Topology. Proceedings, 1986–87. VI, 366 pages. 1989.

Vol. 1370: G. Carlsson, R.L. Cohen, H.R. Miller, D.C. Ravenel (Eds.), Algebraic Topology. Proceedings, 1986. IX, 456 pages. 1989.

Vol. 1371: S. Glaz, Commutative Coherent Rings. XI, 347 pages. 1989.

Vol. 1372: J. Azéma, P.A. Meyer, M. Yor (Eds.), Séminaire de Probabilités XXIII. Proceedings. IV, 583 pages. 1989.

Vol. 1373: G. Benkart, J.M. Osborn (Eds.), Lie Algebras, Madison 1987. Proceedings. V, 145 pages. 1989.

Vol. 1374: R.C. Kirby, The Topology of 4-Manifolds. VI, 108 pages. 1989.

Vol. 1375: K. Kawakubo (Ed.), Transformation Groups. Proceedings, 1987. VIII, 394 pages, 1989.

Vol. 1376: J. Lindenstrauss, V.D. Milman (Eds.), Geometric Aspects of Functional Analysis. Seminar (GAFA) 1987–88. VII, 288 pages. 1989.

Vol. 1377: J.F. Pierce, Singularity Theory, Rod Theory, and Symmetry Breaking Loads. IV, 177 pages. 1989.

Vol. 1378: R.S. Rumely, Capacity Theory on Algebraic Curves. III, 437 pages. 1989.

Vol. 1379: H. Heyer (Ed.), Probability Measures on Groups I. Proceedings, 1988. VIII, 437 pages. 1989